Efficiency in Environmental Regulation

Studies in Risk and Uncertainty

Editor:
W. Kip Viscusi
Department of Economics
Duke University
Durham, North Carolina 27706, U.S.A.

Efficiency in Environmental Regulation

A Benefit-Cost Analysis of Alternative Approaches

Ralph A. Luken
U.S. Environmental Protection Agency
Washington, D.C.

Kluwer Academic Publishers
Boston/Dordrecht/London

Distributors for North America:
Kluwer Academic Publishers
101 Philip Drive
Assinippi Park
Norwell, Massachusetts 02061 USA

Distributors for all other countries:
Kluwer Academic Publishers Group
Distribution Centre
Post Office Box 322
3300 AH Dordrecht, THE NETHERLANDS

Library of Congress Cataloging-in-Publication Data

Luken, Ralph Andrew.
    Efficiency in environmental regulation : a benefit-cost analysis
of alternative approaches / Ralph A. Luken.
        p.   cm. — (Studies in risk and uncertainty)
    Includes bibliographical references (p.   ).
    1. Paper industry—Environmental aspects—Government policy-
-United States.   2. Wood-pulp industry—Environmental aspects-
-Government policy—United States.   3. Environmental law—United
States.   I. Title.   II. Series.
HD9826.L84   1990
363.73´ 1—dc20
                                                            90–4565
                                                            CIP

7-29-92

CONTENTS

TABLES

FIGURES

FOREWORD

A management agency -- such as a publicly or privately owned electric utility -- must, if it is to be efficient in carrying out its day-to-day tasks, have a means of monitoring its performance to assess the efficiency of its operations and the effectiveness of its planning. For example, how did the demand for electricity compare with that assumed in planning? How effective were the incentives applied to induce energy conservation by users? Such ex post analyses are essential for improving the planning process and hence for improving decisions with respect to efficiency and resource allocation.

Unfortunately, it seems to be very difficult for public agencies to make such ex post evaluations an integral part of agency activities, whether the agencies are "producers," e.g., the Corps of Engineers or the Bureau of Reclamation with respect to water resources management, or are regulatory agencies such as the Environmental Protection Agency or the Food and Drug Administration. Here and there a few ex post analyses of agency programs have been done, but rarely by the responsible agency itself. These analyses have attempted to compare the results actually achieved with the results estimated in planning, either in terms of project outputs or in terms of effectiveness of regulatory and/or economic incentives in inducing desired changes in behavior. For example, Robert Haveman of the University of Wisconsin did an ex post analysis of flood damage reduction benefits of Corps of Engineers projects, comparing the mean annual benefits realized with the mean annual benefits estimated in planning. The Economic Research Service of the Department of Agriculture analyzed at least a half dozen P.L. 566 projects, some years after the projects were initiated, to compare the time stream of actions realized with the time stream estimated in planning.

With respect to the effectiveness of incentives, James Hudson, in *Pollution-Pricing* [1981] examined the response to sewer surcharges in five major U.S. cities. The Pacific Northwest Power Council, in conjunction with the various utilities in its area, including Bonneville Power Administration, analyzed the effectiveness of the set of incentives developed jointly, and subsequently applied, to induce residential, commercial, industrial, and agricultural energy users to adopt energy conservation. The most publicized topic for ex post analysis has been deregulation of the airline industry. Multiple reports by various protagonists have been published on that topic.

The Environmental Protection Agency has wide-ranging responsibilities, e.g., air quality, water quality, hazardous waste disposal, and pesticides management. In carrying out these responsibilities, massive numbers of regulations have been issued. Presumably the regulations have been designed to induce changes in behavior so that discharges to the environmental media will be reduced, in order that (1) ambient environmental quality will be improved, (2) resulting in benefits to humans and other species, (3) at a cost that is less than the value of these benefits. The implicit assumption of environmental legislation has been that this sequence of results would occur as a result of promulgated regulations.

One looks in vain for the ex post analyses of EPA actions which demonstrate the chain of relationships noted above. Ambient water quality in relation to various water quality indicators has been tabulated for many of the nation's river reaches at several points in time; the same has been done for ambient air quality in various metropolitan areas. Little attempt has been made, however, to analyze specifically (1) how the positive changes in ambient quality that have occurred actually have been achieved (i.e., did the improvements result from reductions in discharges from operating plants or simply as a result of many such plants being closed), (2) what increases in benefits can be attributed to the improvements in ambient quality, and (3) does the value of the benefits achieved exceed to costs incurred.

Some aggregate benefit-cost analyses of water and air regulations have been done, primarily by A. Myrick Freeman III in 1982. But these suffer from the same limitations as would exist if, for example, one did a water supply-demand analysis for the U.S. as a whole. Such analysis would show that there is plenty of water in the aggregate, but would completely mask various regional problems. To define problems and to determine the efficiency of environmental regulations require analyses at the local level and in terms of specific river reaches and airsheds. It is this level of analysis that makes Skip Luken's work a significant contribution and a step forward.

By focussing on the pulp and paper industry, he is able to evaluate three approaches to regulation used by EPA in terms of the relationship between benefits achieved and costs incurred. To do this for an industry as complex as the pulp and paper industry is a major task. No one who knows anything about the industry would claim that the cost numbers are "exact"; too many site-specific variables affect costs. Similarly, multiple factors affect benefits associated with a given river reach or a given urban area. Nevertheless, as the sensitivity analysis shows, useful insights into the efficiency of regulations are gained even considering ranges in the estimates. The results suggest that a regional or subregional approach to environmental regulation is preferable to a national approach, in terms of efficiency.

Skip Luken is to be commended for tackling a difficult but much needed analysis. One can only hope that such ex post analyses would become integral activities of governmental agencies at all levels, and that the results would be used in developing and implementing more efficient regulatory programs.

Blair T. Bower, P.E.

PREFACE

This book is based upon a variety of benefit-cost analyses undertaken by the U.S. Environmental Protection Agency (EPA). Some of these analyses were performed as part of EPA's regulatory development efforts for its regulations of the pulp and paper industry. Others were originated as part of EPA's research into the benefits of its regulatory programs and the alternative methodologies for measuring and evaluating those benefits.

While the research is EPA's, the decision to publish this assessment of the efficiency of environmental regulations is my own and, I must emphasize, the interpretations and opinions presented are my own and do not necessarily reflect the views of EPA.

This work is the first systematic economic efficiency assessment of the United States' experience in environmental quality management. I hope that the lessons learned, both in development of assessment methods and in the conclusions about the relative efficiency of alternative regulatory approaches, will contribute to the public and professional debate about the future of the United States' environmental protection program.

The core of this book owes much to the team that prepared the original benefit-cost analyses of pulp and paper industry regulations for EPA. The costs and residuals models for water and air were developed by Tom Birdsall and Ken Wise of Putnam, Hayes and Bartlett, Inc. and Jim Wilson of E.H. Pechan and Associates. Walter Grayman, a consulting engineer, prepared the water quality analysis and Chris Maxwell of Versar Inc. prepared the air quality analysis. Marcus Duff of Mathtech prepared the particulate matter benefit analysis that was originally developed for the Regulatory Impact Analysis for the particulate matter ambient air quality standard. Much of the other analysis is based on consultant studies prepared for EPA in support of regulation development. Reed Johnson, a part-time consultant to the Economic Analysis Branch, assisted in the water quality benefit analysis that was part of the preliminary Regulatory Impact Analysis for the Best Conventional Technology effluent guideline. Greg Michaels' work on the toxicity characteristic rulemaking contributed to the analysis on toxic pollutants. Harry Bondareff, Elaine Danyluk, Virginia Kibler, and Margaret Miller assisted in preparation of the benefit-cost analyses among their other duties as research assistants in the Economic Analysis Branch. Blair Bower, a part-time consultant to EPA, added significantly to the direction and rigor of the analysis. Extensive discussions with Lyman Clark, head of the Economic Impact Analysis Branch in OPPE in the early 1970s, reformulated the evaluation into the critique of alternative approaches to environmental regulation and played a major role in the quality of the final document. Robert Anderson, my predecessor in the Benefits Staff, added to the clarity of the presentation. Anne Grambsch improved the quality of the draft with thoughtful comments. Peggy Miller provided valuable editorial assistance.

During the five-year period over which the case study and then the critique of environmental regulations emerged, I only found time to work on the project at nights and on weekends. I want to thank my wife for her tolerance and her warm welcome at home and at our retreat in Nanjemoy, MD. I also want to acknowledge the understanding of my friends in the Washington community who will certainly be relieved not to hear about this project any more.

I would like to dedicate this book to Lyle Craine and Blair Bower. Lyle was my professor and associate in the School of Natural Resources at the University of Michigan in the 1960s. He taught me about the importance of ex post evaluations as a way to gain insights into the design of effective government programs. Blair was a major participant in the Quality of the Environment Program at Resources for the Future. He taught me about the practical techniques for conducting ex post evaluations of environmental programs. I hope this effort captures some of what they thought was important.

Ralph A. "Skip" Luken

Efficiency in Environmental Regulation

Chapter 1

INTRODUCTION

The newly created U.S. Environmental Protection Agency (EPA) in the 1970s embarked on three different regulatory approaches for reducing pollution. The Clean Air Act of 1970 mandated that EPA set ambient-based standards that protect human health and welfare. These standards require existing industrial sources of air pollution to install pollution control equipment only to the extent necessary to meet ambient standards. The Clean Air Act also mandated that EPA in some circumstances set benefits-based[1] standards that trade off risks to society with the costs of risk reduction. These standards require existing industrial sources to install pollution control equipment only to the extent that there would be a reasonable balance between the benefits of pollution reduction and the costs of pollution control technology. The Clean Water Act of 1972, however, mandated that EPA set technology-based standards that reflect the availability and affordability of pollution control technology. These standards require existing industrial and municipal sources to meet uniform discharge limitations, even if the pollutants discharged did not result in violations of ambient standards.

A reading of the *Congressional Record* does not provide a clear picture of why the United States shifted from predominantly ambient-based standards to technology-based standards [Library of Congress. 1973]. Part of the answer lies in earlier Clean Air Act and Clean Water Act legislation. The Clean Air Act of 1967 focused primarily on technical feasibility, but that approach apparently did not generate an adequate basis for setting emission limitations. The Clean Water Act of 1965 focused on nationally required, though seldom established, ambient water quality standards for interstate waters, but that approach apparently resulted in insufficient effluent reduction.

Another part of the answer lies in the perceived absence, in 1972, of adequate models for tracing the effects of pollutants on ambient water quality. If this were the case, one would have thought the even greater difficulties in air quality modeling would have caused Congress to hesitate in adopting an ambient-based approach for air pollution in 1970. Instead, Congress adopted the ambient approach for most of its air pollution standards in 1970, but switched to the technology approach for water standards in 1972.

At the time of the shift from an ambient to a technology approach for the water program, EPA Administrator William Ruckelshaus testified against basing effluent limitations entirely on the availability of technology. The Administrator warned:

> In completely obliterating the admittedly complex relationship between municipal and industrial effluent and water quality, the Senate may have sacrificed wisdom for simplicity. It is simpler to address the problem in terms of available technology than in terms of complex treatment and ambient water quality relationships. But it is my conviction that in the environment as elsewhere, the renunciation of known complexity on the altar of simplicity is the essence of bad government policy [Library of Congress. 1973].

Also, several research studies before and after the early 1970s concluded that technological dominance of regulatory programs is achieved at a high social cost and with poorer environmental results than anticipated.[2]

Luken and Pechan [1977] prepared for the National Commission on Water Quality, the only national, geographically specific evaluation of the new technology approach to water pollution control. They assessed the impacts between 1975 and 1985 of population and economic growth on water quality in the United States, subdivided into 100 river basins. They concluded that uniform implementation of the technology-based standards required by the Clean Water Act of 1972 would be costly in relationship to the improvements in water quality and distract attention from non-point source[3] contributions to water quality problems. They did not assess alternative regulatory approaches, but did suggest limiting the implementation of technology-based standards to those basins where there were severe water quality problems and where reductions in pollutants from point sources would contribute to improvements in water quality.

To the degree that EPA's regulatory approaches are efficient, society receives net benefits from the reduction or elimination of environmental damages. Society pays for pollution reductions through increased costs for goods and services provided by polluting entities. Society receives net benefits whenever the societal benefits of better health or a cleaner environment have greater value than the costs of the pollution reductions [Gramlich. 1981, Stokey and Zeckhauser. 1978].

Freeman [1982] has attempted the only national ex post assessment of the benefits and costs of the U.S. water and air pollution control programs. Portney and Freeman [1990] recently reiterated the same findings with an update of Freeman's original effort. In both publications, the findings suggest that the benefits have exceeded the costs for ambient-based standards associated with the Clean Air Act and that the costs have exceeded the benefits for the technology-based standards associated with the Clean Water Act. As both Portney and Freeman acknowledge, their estimates are not systematic, spatially-specific assessments of the efficiency consequences of environmental regulations. As Blair Bower pointed out in the forward to this book, these estimates, although insightful, suffer from the same limitations that would exist if, for example, one did a water supply-water demand analysis for the United States as a whole. Such an analysis would show that there is plenty of water in the aggregate, but would completely mask various regional problems. A reasonably accurate assessment of the net benefits of environmental regulations requires systematic analyses at many locales and in terms of specific river reaches and airsheds.

To date, there have been no national, geographically specific evaluations of the success of EPA's regulatory approaches in generating net benefits. Such an assessment of net benefits would require information about four distinct relationships [Freeman. 1982, Portney. 1990]. The first relationship characterizes the regulatory-induced reductions in pollutant discharges that result from investments in process changes and pollution control technology. The second relationship characterizes the spatial and temporal changes in environmental quality that result from spatial and temporal changes in pollutant discharges. The third relationship characterizes the physical protection of humans and other species that results from an improvement in ambient environmental quality. The physical protection could be reductions in risks to human health and natural and man-made environments. Or, it could be altered patterns of recreation activity. The fourth relationship characterizes the changes in utility or welfare that result from environmental protection. These changes can, to varying degrees, be measured in monetary terms i.e., individuals' willingness-to-pay. An estimate of net benefits results from comparing the benefits of improved environmental quality with the costs of pollutant reduction. These estimates usually compare steady-state annual benefits and annual costs at some specified point in time.

This book is the first effort to compare EPA's actual success in generating net benefits through the three different regulatory approaches - - i.e., the technology approach (the water program), the ambient approach (the ambient-based air program), and the benefits approach (the benefits-based air program). It compares the measurable environmental benefits and the costs that result from applying technology, ambient, and benefits approaches to one major industry -- the pulp and paper industry.

Assessing the success of EPA's different regulatory approaches on the basis of generating net benefits is an important evaluation criterion, but should not be the only evaluation criterion. A more comprehensive measure of success would be reducing unreasonable risks as defined in a proposed organic statute for EPA [CF. 1988]. An evaluation of success on the basis of reducing unreasonable risk would include at least the following criteria: individual and population risks, the distribution of benefits and costs, and the net benefits. No such evaluation of EPA regulatory approaches has ever been done because of the difficulty of quantifying the other criteria and integrating them into a composite measure of success. Nonetheless, these other criteria are important in evaluating EPA's regulatory approaches. If EPA's regulatory programs do not result in net benefits, the values of these other criteria should be sufficiently great to offset the loss of societal resources that could be directed to reducing environmental risks.

The book focuses on the pulp and paper industry because the industry is the largest industrial user of water (which is the best available proxy for conventional water pollutants) and a significant source of air pollutants, and because the industry is subject to the three different regulatory approaches. Equally important in the choice is the fact that the production processes and conventional pollutant loadings of the pulp and paper industry are reasonably well documented and understood. Appendix A to this book provides a brief description for the reader unfamiliar with the industry of its standard manufacturing processes and the kinds of environmental pollutants commonly associated with these processes.

The book assesses the results of EPA's regulation of conventional air and water pollutants from the pulp and paper industry between 1973 and 1984. It also examines the potential of the three different regulatory approaches for achieving net benefits from air, water, and hazardous waste pollution control between 1984 and 1994. Such an examination is necessary because of the continuing triumph of the technology approach over the ambient approach and because of the total neglect of the benefits approach in EPA regulations except in the regulation of chemicals and some air pollutants [Bonnie. 1975]. The Clean Air Act Amendments of 1977 and Clean Water Act Amendments of 1977 gave additional impetus to the technology approach for existing industrial sources of pollution [LaPierre. 1977, Randle. 1979]. More recently, the Resource Conservation and Recovery Act of 1984, the Safe Drinking Water Act of 1986, and the proposed amendments in 1989 to the Clean Air Act for hazardous air pollutant control represent the new triumphs of the technology approach over the ambient and benefits approaches.

Notes

1. In this book we use the term "benefits-based" to refer to standards that are established by comparing potential benefits with potential costs. These standards might more precisely be referred to as "net-benefits-based" standards. The term "benefits-based" is used for the sake of simplicity and readability, however, and should be interpreted as implying an evaluation of net benefits.

2. For a summary of these studies, see Tietenberg [1985].

3. "Non-point" sources are those which do not flow from a discrete point, such as a discharge pipe. Non-point sources include urban and agricultural runoff and storm water overflow.

References

Bonnie, John E. 1975. "The Evolution of 'Technology-Forcing' in the Clean Air Act." *Environment Report*, 6:1-30.

The Conservation Foundation (CF). 1988. "The Environmental Protection Act," Second Draft. Washington, DC.

Freeman, A. Myrick III. 1982. *Air and Water Pollution Control: A Benefit-Cost Assessment*. John Wiley and Sons, 1982.

_____. 1990. "Water Pollution Policy" in *Public Policies for Environmental Protection*. Paul R. Portney, editor. Resources for the Future, Inc., Washington, DC.

Gramlich, Edward M. 1981. *Benefit-Cost Analysis of Government Programs*. Prentice Hall, Englewood Cliffs, NJ.

LaPierre, Bruce D. 1977. "Technology-Forcing and Federal
Environmental Protection Statutes." *Iowa Law Review*, 62:771-908.

Library of Congress. 1973. "A Legislative History of the Water Pollution
Control Act Amendments of 1972." U.S. Government Printing Office,
Washington, DC.

Luken, Ralph A. and Edward H. Pechan. 1977. *Water Pollution Control:
Assessing the Impacts and Costs of Environmental Standards*. Praeger,
New York, NY.

Portney, Paul R. 1990. "Air Pollution Policy" in *Public Policies for
Environmental Protection*. Paul R. Portney, editor. Resources for the
Future, Inc., Washington, DC.

Randle, Russell V. 1979. "Forcing Technology: The Clean Air Act
Experience." *The Yale Law Journal*, 88:1713-1739.

Stokey, Edith, and Richard Zeckhauser. 1978. *A Primer for Policy
Analysis*. Norton, New York, NY.

Tietenberg, Thomas H. 1985. *Emissions Trading: An Exercise in
Reforming Pollution Policy*. Resources for the Future, Inc., Washington,
DC.

Chapter 2

REGULATORY APPROACHES

EPA's central mission is to carry out its various statutory directives to protect the nation's health, welfare, and environment from risks posed by pollution (Table 2-1). Statutes enacted by Congress can be quite specific, but all require clarification or refinement to make the regulatory program operational. Regulations establish the particular policies and procedures that EPA uses to carry out its statutory goals.

Table 2-1. EPA's Major Statutes

Statute	Date Amended
Clean Water Act (CWA)	1948, 1956, 1972, 1977 and 1987
Clean Air Act (CAA)	1965, 1970 and 1977
Comprehensive Environmental Response, Compensation and Liability Act (CERCLA)	1980 and 1986
Federal Insecticide, Fungicide and Rodenticide Act (FIFRA)	1976 and 1986
Resource Conservation and Recovery Act (RCRA)	1965, 1970, 1976 and 1984
Safe Drinking Water Act (SDWA)	1974, 1979, 1980 and 1986
Toxic Substances Control Act (TSCA)	1976 and 1986

This chapter briefly describes the elements of a regulatory program and then discusses EPA's regulatory programs under the Clean Water Act, Clean Air Act, and the Resource Conservation and Recovery Act as they affect the pulp and paper industry. We emphasize the Clean Water and Clean Air Acts because standards set under these two acts essentially define the differences between technology, ambient, and benefits regulatory approaches.

ELEMENTS OF A REGULATORY PROGRAM

A regulatory program consists of four activities -- standards, permits, compliance monitoring, and enforcement. The major purpose of these four activities is to encourage, guide, or prohibit future conduct that would be detrimental to the environment.

Developing Standards

The first essential element in most of EPA's regulatory programs is to issue standards that clarify EPA's approach to mitigating environmental problems [McGarity. 1983]. Technology-based standards specify a discharge limitation based on the availability and affordability of technology, without consideration of the quality of the ambient environment. Ambient-based standards specify a safe or clean environment based on protection of human health and welfare, without consideration of cost or technological feasibility. Between these two extremes, benefits-based standards focus on the trade-off between the risks to society from pollution and the costs society must pay to reduce pollution.

Ambient- and technology-based standards are the predominant standards that most directly affect the pulp and paper industry. EPA relies primarily on technology-based standards to carry out its mission under the Clean Water Act and the Resource Conservation and Recovery Act and primarily on ambient-based standards to carry out its mission under the Clean Air Act. EPA uses benefits-based standards to a limited extent under the Clean Air Act.

Issuing Permits

The second essential element in most of EPA's regulatory programs is to ensure that polluting facilities obtain operating permits. Permits tailor national standards to individual facilities and their environmental circumstances. State governments administer most permit programs. In some circumstances, these permit programs are delegated to the states by EPA, and EPA sets requirements for an approvable permit program. States run the majority of the water permit programs, most (along with some local governments) of the air permit programs, but few of the hazardous waste permit programs. State permit programs for criteria air pollutants from existing sources and for solid wastes do not require federal approval.

Monitoring Compliance

The third element in most of EPA's regulatory programs is to monitor the compliance of facilities with the conditions of their permits. There are four sources of compliance information: (1) self-monitoring and reporting by the source of pollutants, (2) inspections by government officials or independent third parties, (3) citizen complaints, and (4) ambient monitoring. The mix of information sources used varies from one environmental program to another, but self-monitoring and inspections are by far the most important means of compliance monitoring.

Enforcing Regulations

The fourth element in most of EPA's regulatory programs is to provide a wide range of enforcement responses to violations of environmental operating permits. Traditional enforcement sanctions fall into four general categories (informal, administrative, civil, and criminal). A variety of informal responses, such as warning letters and phone calls, fall at one end of the enforcement spectrum. Beyond these informal responses, EPA and state agencies use administrative, civil, and criminal remedies and sanctions.

THE CLEAN WATER ACT

Through the 1950s and 1960s, the national approach to water pollution deferred to states to set ambient water quality standards and to develop plans to achieve these standards. The Clean Water Act of 1972 significantly changed this approach. It combined new ambient-based and technology-based standards, with a strong emphasis on the latter [Baum. 1983]. The amendments called for compliance with the technology-based standards by specific dates and created a strong federal enforcement program. The Clean Water Act was amended in 1977 and 1987.

Technology-Based Standards

The Clean Water Act calls on EPA to set technology-based standards that restrict the amount of pollutants discharged by industrial plants. The standards are different for each industry and for subcategories within an industry. Three technology-based standards apply to existing industrial sources: Best Practicable Control Technology, Best Conventional Technology, and Best Available Technology Economically Achievable. New Source Performance Standards, which apply to new sources, are similar to Best Available Technology.

The development of technology-based standards (called effluent guidelines by EPA) follows four distinct stages: (1) the contractor study stage, (2) the proposed guideline stage, (3) the promulgated rule stage, and (4) the judicial review stage [Magat, Krupnick, and Harrington. 1986]. The judicial review stage was particularly important in developing the Best Conventional Technology standards and partly explains why their development took almost 10 years.

To develop the technology-based standards for the pulp and paper industry, EPA initiated contractor collection of data on the industry in 1972. EPA studied the pulp, paper, and paperboard industry to determine whether differences in raw materials, final products, manufacturing processes, equipment, age and size of facilities, water use, wastewater constituents, or other factors required the development of separate regulations for different segments of the industry. Based on the results of the data-gathering program, EPA divided the pulp and paper industry into three major segments: (1) integrated mills where pulp, pulp and paper, or pulp and paperboard are produced, (2) secondary fiber mills where paper or paperboard is made from wastepaper, and (3) nonintegrated mills where paper or paperboard is manufactured, but pulp is not produced on site.[1] It further divided the segments into 25 subcategories based on the varying nature of processes that were employed and products that were manufactured.

EPA identified several control and treatment technologies, including both internal and end-of-pipe methods, that are in use or are capable of being used, to reduce the generation of or to treat pulp, paper, and paperboard wastewaters. For each technology, EPA did the following:

(1) compiled all available data, (2) analyzed the capabilities of the internal modifications and of the treatment systems, (3) considered the non-water quality environmental impacts, such as impacts on air quality, solid waste generation, and energy requirements, and (4) estimated the costs and economic impacts of applying the technology.

EPA promulgated Best Practicable Technology standards to control the familiar, primarily conventional, pollutants for most of the pulp and paper industry in 1974 and 1977 [FR. 1974, 1977] and promulgated the remaining Best Practicable Technology and New Source Performance Standards for conventional pollutants in 1982 [FR. 1982]. The Best Practicable Technology standards for Biological Oxygen Demand (BOD) and Total Suspended Solids (TSS) are based on biological treatment for all but a few nonintegrated subcategories of the pulp and paper industry where only primary treatment is the basis for the effluent limitations. These limitations for BOD and TSS are uniquely specified for the 25 production/process subcategories within the industry. The New Source Performance Standards for conventional pollutants were established based on the same technology as Best Practicable Technology, plus additional flow-reducing, in-plant process controls.

EPA promulgated the Best Conventional Technology standards in December 1986 [FR. 1986(b)]. Because of the high costs of additional treatment technology, EPA set these effluent limitations for additional conventional pollutant control equal to Best Practicable Technology effluent limitations for all subcategories.

The recent work of Magat, Krupnick, and Harrington [1986] is the only systematic effort to explain the implicit decision rules that EPA used to set Best Practicable Technology effluent standards for industrial water polluters.[2] Their review of effluent standards for several industrial categories revealed drastic differences in the stringency of the final standards based on the cost per kilogram of BOD removed and in changes in stringency of the standard during the different stages of regulation development. They applied statistical analysis of the standard-setting process for 106 subcategories, including five subcategories of the pulp and paper industry, to explain the variations in stringency among industries and over time. They found that differences in compliance costs, the quality of information generated by the contractor reports, industry characteristics (trade association budgets, industry concentration, and profitability), and formal industry comments influenced the degree of stringency. Contrary to the traditional wisdom, economic impact projections (plant closings, unemployment, and price increases) did not significantly affect the stringency of the standards.

Technology-Based Toxics Standards

The 1977 amendments to the Clean Water Act defined 129 pollutants as toxic. Existing sources that discharge those pollutants were required to meet Best Available Technology criteria by July 1984, and new sources were required to meet New Source Performance Standards criteria when they commenced operation. After issuing phase two Best Practicable Technology standards in 1977, EPA's regulatory effort concentrated on acquiring data on the presence of the 129 priority and toxic pollutants in industrial wastewaters, including those of the pulp and paper industry. EPA promulgated the final Best Available Technology and New Source Performance Standards for the pulp and paper industry in 1982 [FR. 1982]. EPA established technology-based limitations for three pollutants -- pentachlorophenol, trichlorophenol, and zinc.

We do not estimate the costs or the effluent reductions of toxics required to comply with Best Available Technology effluent guidelines because EPA estimated that the Best Available Technology guidelines would not require any additional pollution control expenditures for mills existing in 1984 and only minor costs compared to conventional pollution control costs for expanded or new mills between 1984 and 1994 [EPA. 1982]. Best Available Technology consists of costless chemical substitution for pentachlorophenol and trichlorophenol and minor costs for zinc removal, which are incorporated in the costs to comply with the Best Practicable Technology effluent guidelines.

Around the same time that EPA issued its Best Available Technology limitations for toxic water pollutants from pulp and paper mills, evidence began to emerge that these mills might be the source of another toxic pollutant, dioxin. Actually, pulp and paper manufacture has been suspected as a source of dioxin pollution since at least 1980 [EPA. 1980]. In 1983, the State of Wisconsin discovered dioxin-contaminated fish at the commercial carp fishery in the Petenwell Flowage Reservoir. The reservoir is downstream from several pulp and paper mills on the Wisconsin River. The state suspected that the dioxin came from the banned trichlorophenol that had been used as a biocide and closed the carp fishery the same year [EPA. 1987].

In 1983, Congress passed a special appropriation for a National Dioxin Study by EPA to determine potential sources of and the presence of dioxin in the environment. The EPA study established seven site categories (tiers) for dioxin sampling, ranging from the most probable tier of dioxin contamination to the least probable.[3]

As part of its tier seven least probable background investigation, EPA collected whole fish and sludge samples around 395 sites [EPA. 1987]. Of these 395 study sites, EPA found detectable levels of dioxin at 112 sites (28 percent) and detected dioxin at greater than five parts-per-trillion (ppt) at 38 sites (10 percent). For the 64 pulp mill sites in the study, however, EPA found detectable levels of dioxin at 34 sites (53 percent) and detected dioxin at greater than five ppt at 10 sites (15 percent). At two of these pulp and paper mill sites, EPA found the highest dioxin levels in whole fish (the Androscoggin River, Maine - maximum 29 ppt, and the Rainy River, Minnesota - 85 ppt). EPA also found dioxin in the samples of wastewater treatment sludge from 12 mills with a maximum dioxin level of 414 ppt in the sludge from a bleached kraft mill.

As a result of these findings, EPA agreed to a limited, joint study with the pulp and paper industry in June 1986. The "5-Mill Study" sampled five mills located in Ohio, Oregon, Maine, Minnesota, and Texas between June 1986 and January 1987. Results from the study show dioxin detected in the bleached pulp and sludge from all five mills and in the final effluents from four of the five mills [EPA. 1988].

EPA followed up on the "5-Mill Study" by initiating a more comprehensive characterization of dioxin levels at bleached pulp mills using chlorine or chlorine derivatives. The American Paper Institute (API) agreed to participate in this more comprehensive assessment. In April 1988, EPA and API signed a cooperative agreement to gather data collectively for the "104-Mill Study" [FR. 1988]. EPA anticipates results from the "104-Mill Study" in the spring of 1990.

Technology-Based Standards for Indirect Dischargers

Pulp and paper mills that discharge to city sewers are indirect dischargers, as opposed to direct dischargers to bodies of water. Thus, they do not need a permit under the Clean Water Act, but must comply with EPA pretreatment standards. Pretreatment standards consist of either national standards, which are designed to control toxic discharges from specific industry categories (e.g., electroplating and metal finishing), or indirect discharger standards, which are designed for specific types of municipal wastewater treatment plants to prevent interference with their operation and pass-through of toxic pollutants.

EPA promulgated pretreatment standards for the pulp and paper industry in November 1982. The standards established limits for three toxic pollutants -- pentachlorophenol, trichlorophenol, and zinc. EPA required indirect-discharging pulp and paper mills to be in compliance with those standards in three years.

EPA regulations required municipalities with indirect dischargers to establish pretreatment programs that include user charges by July 1983. The pretreatment programs for most pulp and paper mills require installation of internal controls and payment of user charges to municipalities based on the volume and strength of conventional pollutants.

We estimate the costs and effluent reductions of conventional pollutants to comply with municipal pretreatment programs in Chapter 3. We could not assess the environmental benefits resulting from the compliance of pulp and paper mills with municipal pretreatment programs, however, because of the impossibility of isolating the impact of the wastes from pulp and paper mills on water quality. Moreover, neither EPA nor the states have any systematic data on the actual or estimated reductions of conventional pollutants that have resulted from industry compliance with municipal pretreatment programs.

THE CLEAN AIR ACT

Through the 1950s and 1960s, the federal regulatory strategy deferred to state governments. The federal initiatives focused on research and financial and technical support of state programs. The Clean Air Act of 1970 significantly altered the national approach to air pollution. It called for the federal government to establish national ambient air quality standards and for states to implement them. It also required the federal government to set technology-based standards for new industrial sources and for hazardous air pollutants, and, in a few cases, to set benefits-based standards for existing sources. The Clean Air Act was amended in 1977.

Ambient-Based Standards

The Clean Air Act calls on EPA to set primary and secondary National Ambient Air Quality Standards. Primary standards are intended to protect the health of the population, whereas secondary standards are meant to protect public welfare (materials, crops, forests, etc.).

The development of ambient air quality standards follows five distinct stages: (1) criteria document stage -- summary of the latest scientific knowledge, (2) staff paper -- interpretation of scientific data, (3) proposal stage -- a draft ambient standard, (4) promulgation stage -- a final ambient standard, and (5) judicial review -- litigation, settlements, and revisions [Jordan. 1983]. The development or revision of an ambient air quality standard takes between five and ten years.

The Clean Air Act requires EPA to develop for each criteria pollutant "air quality criteria" that summarize the latest state of scientific knowledge concerning the pollutant and its effects on human health and the environment. Thus, the standard-setting process begins with an extensive review of all the scientific information concerning a pollutant, including reported health and welfare effects, sources of the pollutant, and atmospheric chemistry. The criteria for each pollutant are published in a "criteria document," which critically reviews the health and welfare effects evidence from a scientific point of view.

As soon as the criteria document appears to be substantially complete, EPA begins preparing a "staff paper," which reflects the staff's interpretation of the key studies and scientific evidence described in the criteria document and identifies critical elements to be addressed in the standard-setting process. The staff paper helps to bridge the gap between the scientific information contained in the criteria document and the judgments required of the Administrator in setting the ambient standards. Depending upon the complexity of the standard under review, the criteria document and staff paper require two to three years to complete, including review by the Clean Air Scientific Advisory Committee.

When the Administrator reaches a decision on the standard, the regulation is proposed by EPA in the *Federal Register*, and public comments are solicited. One or more public meetings are generally held after the proposal to provide additional opportunity for public comment. After reviewing all the public comments and assessing any new or additional information submitted during the public comment period, a final regulatory package is developed, reviewed in accordance with the previously described procedure, and promulgated.

In response to the Clean Air Act, EPA has established air quality standards for six pollutants: carbon monoxide (CO), lead (Pb), nitrogen dioxide (NO_2), ozone (O_3), particulate matter (PM), and sulfur dioxide (SO_2). The standards that most directly impinge upon the pulp and paper industry are those for particulate matter (total suspended particulates (TSP) in 1971 and particulate matter less than 10 microns in diameter (PM10) in 1987) and sulfur dioxide.

To develop the TSP national ambient air quality standard required by the 1970 Clean Air Act, the newly created EPA turned to the efforts of one of its predecessors, the National Air Pollution Control Administration.   The 1967 Clean Air Act required the National Air Pollution Control Administration to develop and issue "criteria of air quality" and for states to adopt ambient air standards consistent with the criteria. The Administration issued a particulate matter criteria document in 1969 [HEW. 1969].

The 1970 Clean Air Act required the Administrator of EPA to propose national ambient air quality standards 30 days after the passage of the Act. According to Marcus [1980], Administrator Ruckelshaus was committed to meeting this statutory deadline and turned to officials from the old National Air Pollution Control Administration, which was incorporated into EPA, for a full set of ambient standards. The officials gave him the standards, using scientific evidence and judgments that they had formed about the effects of air pollutants, three days before the statutory deadline.[4]   Administrator Ruckelshaus had neither time nor independent staff to review these standards, so he felt he had no choice but to accept the recommendations of these officials. The Administrator proposed national ambient air quality standards for six criteria pollutants, including TSP, on January 30, 1971, and promulgated final standards on April 30, 1971 [FR. 1971].[5]

Benefits-Based Standards

The Clean Air Act requires EPA to set New Source Performance Standards for new or modified stationary source categories whose emissions cause or significantly contribute to air pollution that may endanger public health or welfare. New Source Performance Standards are to be based on the best demonstrated technology, considering costs and economic and social impacts.

If a New Source Performance Standard is set for a pollutant not regulated by an ambient standard, the Clean Air Act requires that states develop regulations to control existing sources of the same pollutant. EPA regulations specify, however, that EPA will issue guidance to the states on the best retrofit technology for existing sources, considering costs and economic and social impacts. If the pollutant is health-related (e.g., sulfuric acid mist from sulfuric acid plants), states must provide strong justification for adopting emission limits less stringent than the guidance specified by EPA. For pollutants that are welfare related (e.g., total reduced sulfur [TRS] from pulp and paper mills), states have more flexibility in adopting regulations for existing sources. Under these circumstances, the standard becomes a benefits-based rather than a technology-based standard.

The development of a New Source Performance Standard requires at least three years and involves a rigorous engineering and economic investigation, followed by comprehensive EPA and public review [Pahl. 1983]. The technical investigation begins with a nationwide survey of the different processes that are used by the affected industry. These data, together with detailed cost information, are compiled in a background information document.

A number of regulatory alternatives are prepared based on this document. These alternatives represent the spectrum of policy options available for regulating the affected facilities. Generally, each alternative contains a different control technology or group of technologies. The cost and the economic, environmental, and energy impacts of each alternative are then examined and documented. The economic parameters that are critical to determining the reasonableness of a particular alternative are profitability, capital availability, employment, and international trade. After evaluating the results of this analysis, EPA selects one alternative as the basis for the standard. This alternative may include one or more Best Demonstrated Technologies. An emission limit is then selected to reflect the performance of these technologies.

EPA promulgated a New Source Performance Standard for TRS from new sources in 1978 [FR. 1978].[6] The standard reflected EPA's review of the Best Demonstrated Technologies, tempered by EPA's consideration of technology costs and economic impacts. Then EPA, as required by the Clean Air Act, issued guidance for reducing TRS emissions from existing sources in 1979 [FR. 1979(a)]. The TRS emission limits in the emission guidelines were less stringent than the mandatory technology standards for new sources.

We used the TRS emission guidelines for existing sources as an example of a benefits-based standard for two reasons. First, and most important, EPA allowed states considerable flexibility in requiring controls on TRS emissions by its decision to designate TRS as a welfare-based, rather than as a health-based, pollutant [EPA. 1979]. For welfare-related pollutants, states could balance required emission reductions and time required for compliance against other factors of public concern. For example, states took into account the severity of the odor nuisance and the economic conditions of mills in requiring reductions in TRS emissions. The evidence on odor problems around pulp mills in the mid-1980s suggests that states used this opportunity. More extensive guidance from EPA would have undoubtedly contributed to an even better balance between the benefits and costs of odor reductions.

Second, EPA's Office of Air Quality Planning and Standards paid particular attention to cost effectiveness (dollars per ton of pollutant removal) in issuing TRS emission guidelines. It exempted individual sources and precluded control techniques because the cost of control exceeded a certain level. EPA stated, in issuing guidelines for controlling TRS emissions from existing pulp mills:

> The proposed TRS emission limits for new kraft pulp mills are technologically achievable at existing kraft pulp mills when the best control techniques discussed above are applied to each of the eight component process operations. However, the costs of applying the best control techniques are considered excessive for some existing mills, in part because some techniques involve replacement of recovery furnaces or lime kilns. Further, alternative control techniques which are effective but less costly are available for some process operations. Therefore, the cost of applying the various control techniques had a considerable influence on the selection of the recommended best retrofit control technology [EPA. 1979].

Technology-Based Hazardous Air Standards

The Clean Air Act requires EPA to set National Emission Standards for Hazardous Air Pollutants. The Act defines hazardous air pollutants as those for which no air quality standard is applicable, but that are judged to increase mortality or serious irreversible or incapacitating illness. These standards are technology-based and apply to both existing and new stationary sources.

The development of a hazardous air pollutant standard consists of a listing assessment and a source assessment [FR. 1979(b)]. The listing assessment screens potentially hazardous air pollutants to determine their toxicity and potential for public exposure to them. If, after considering comments from the Clean Air Scientific Advisory Committee and the general public, EPA concludes that a substance is or is likely to be either toxic or carcinogenic at ambient levels and that there is significant public exposure to it, the substance will be considered for regulation.

If EPA lists an air pollutant as hazardous, then the Agency conducts a detailed source and exposure assessment. (Some of this assessment may have been done before the listing.) The following factors are considered in regulating a substance: (1) the magnitude of expected risks, (2) the economic and technical feasibility of reducing emissions, (3) the ease of developing and implementing standards for controlling the pollutant, and (4) the impacts of other regulatory efforts, planned or under way. All source categories of the pollutant that are estimated to result in significant risks are evaluated. If a category is already well controlled (for example, by other EPA standards, by other federal, state, or local requirements, or by standard industry practice) and EPA expects that the level of control will continue, then a hazardous air pollutant standard would be redundant and would not need to be established. In that case, a decision not to regulate that source category would be proposed.

If a source category is not already controlled to an acceptable level, then EPA sets an emission standard. In setting the standard, EPA considers risk (both estimated individual lifetime risk and cancer incidences per year), available control techniques, availability of substitute materials, costs, and economic, energy, and environmental impacts. The level of control required may be different for new and existing sources within a source category because of additional problems associated with retrofitting controls in existing sources.

In late 1977, EPA funded a study to screen 632 high-volume industrial organic chemicals for potential evaluation in its hazardous pollutant assessment program [Fuller et al. 1976]. This screening exercise considered both health and exposure factors in determining the priority of substances for assessment. A list of 43 priority compounds was derived from this process. To date, the majority of these compounds have been assessed as potentially hazardous air pollutants.

One of the compounds on this list is chloroform. EPA has assessed air emissions of chloroform from all known sources, including pulp and paper mills. The published conclusions of this assessment were that air emissions of chloroform from all sources can cause significant public health risk, so that the development of emission standards under Section 112 of the Clean Air Act is warranted [FR. 1985].

After publishing this information, EPA investigated individual source categories of chloroform air emissions. To date, the evaluation of chloroform air emissions from pulp and paper mills, potentially the most significant single industrial source of health risks from chloroform, is ongoing. When a decision of whether to develop regulations under the Clean Air Act has been made for this source category, EPA will publish the findings of this analysis in the *Federal Register*.

THE RESOURCE CONSERVATION AND RECOVERY ACT

The Resource Conservation and Recovery Act of 1976 (RCRA) established the most recent federal program regulating solid and hazardous waste management. It defines solid and hazardous wastes, authorizes EPA to set standards for facilities that generate or manage such wastes, and establishes a permit program for hazardous waste treatment, storage, and disposal facilities. It was last amended and reauthorized by the Hazardous and Solid Waste Amendments of 1984. The amendments set deadlines for permit issuance, prohibit the land disposal of many types of hazardous wastes, require the use of specific technologies at land disposal facilities, and establish a new program regulating underground storage tanks.

The term "solid waste" is broadly defined to include discarded materials in liquid, gaseous, or solid form. Congress delegated to EPA the responsibility to redefine "hazardous waste" as solid wastes that may cause or significantly contribute to an increase in mortality or serious illness, or that may cause a substantial hazard to human health or the environment when improperly managed.

Hazardous Wastes

The Resource Conservation and Recovery Act created EPA's hazardous waste management program. It called for EPA to establish criteria for identifying characteristics of hazardous wastes, to issue regulations governing generators and transporters of hazardous wastes, and to establish performance standards for hazardous waste treatment, storage, and disposal facilities.

In May 1980, EPA promulgated its first regulations identifying hazardous wastes. Since then, the regulations have been amended many times, most recently in November 1986. EPA has listed certain wastes as hazardous and has established four characteristics that determine hazardousness -- corrosivity, reactivity, ignitability, and toxicity.

Of most concern to the pulp and paper industry is an EPA-proposed toxicity characteristic regulation [FR. 1986(a)]. It would amend EPA's hazardous waste regulations as follows: (1) by expanding the toxicity characteristic to include 38 additional compounds, including chloroform, (2) by applying compound-specific dilution factors generated from a ground water fate and transport model, and (3) by introducing the Toxicity Characteristic Leaching Procedure. This new procedure would identify wastes that pose a hazard based on their potential to leach significant concentrations of specific toxic species.

The pulp and paper industry produces substantial quantities of solid waste as a consequence of its manufacturing and pollution control processes. The principal waste streams of the industry, however, have not yet been classified as hazardous by EPA. The 1986 proposal could profoundly change the status of certain industry wastes -- especially process wastewaters already regulated under the Clean Water Act -- because it includes chloroform and phenol among the toxicity characteristic contaminants. A more complete description of the potential regulation and its impacts on the pulp and paper industry is provided in Chapter 8.

Notes

1. In the industry lexicon, there are six types of mills that produce pulp and/or paper: (1) the mill produces all the required pulp, which is then made into paper at the mill; (2) the mill produces some pulp and purchases some pulp, which is then made into paper at the mill; (3) the mill produces some pulp which is then used to make paper at the mill, the

rest is sold as market pulp; (4) the mill only produces market pulp; (5) the mill produces paper from secondary fibers (recycled paper); and (6) the mill only produces paper from purchased pulp. Converting operations, i.e., those converting bulk paper into final paper products, occur at paper mills for only a few products.

2. Fraas and Munley [1989] investigated the role of economic criteria in setting the Best Available Technology Economically Achievable standards. They found that the role of cost effectiveness in determining the stringency of industry-specific standards diminished over the period 1981-1986.

3. Unless otherwise indicated, all dioxin results reported from the National Dioxin Study are for 2,3,7,8 - TCDD.

4. The development of the original TSP national ambient air quality standard did not follow the more rigorous and transparent process that characterized the 1987 particulate matter standard. Zaragoza [1982] reviewed the original particulate matter criteria document and found that community observational or epidemiological studies played the major role in setting the primary (health) standards. The 24-hour TSP standard was apparently based on British studies that measured particulate matter with a British Smoke instrument. The background information supporting the proposed TSP standard does not explain how the officials of the National Air Pollution Control Administration combined the morbidity and mortality studies to set the 24-hour standard at 260 ug/m^3, or why they specified that TSP should be measured and not British Smoke. The annual primary standard, 75 μg/m^3, was based on studies that associated increases in the aggravation of bronchitis with an exposure of 80 to 100 μg/m^3 TSP averaged as an annual geometric mean. The secondary (welfare) standard, a 24-hour standard of 150 μg/m^3, was apparently established to protect visibility. In 1979, EPA revised the secondary standard to be equivalent to the primary standard.

5. In 1987, EPA revised the particulate matter ambient air quality standards [FR. 1987]. The revisions changed the focus of the standard from total suspended particulate matter to particulate matter less than 10 microns in diameter.

6. In 1986, EPA issued minor revisions in the New Source Performance Standards for TRS [FR. 1986(c)].

References

Baum, Jonathan K. 1983. "Legislating Cost-Benefit Analysis: The Federal Water Pollution Control Act Experience." *Columbia Journal of Environmental Law*, 9:75-111.

Federal Register (FR). 1971. "Notice of Promulgation of National Primary and Secondary Ambient Air Quality Standards." 36:8186-8201.

_____. 1974. "Pulp, Paper and Paperboard Point Source Category; Effluent Guidelines and Standards." 39:18742-18754.

_____. 1977. "Pulp, Paper and Paperboard Point Source Category; Effluent Guidelines and Standards." 42:1398-1426.

_____. 1978. "Standards of Performance for New Stationary Sources; Kraft Pulp Mills." 43:7568-7578.

_____. 1979(a). "Kraft Pulp Mills; Final Guidelines Document; Availability." 44:29828-29829.

_____. 1979(b). "Policies and Procedures for Identifying, Assessing, and Regulating Airborne Substances Posing a Risk of Cancer." 44:58642-58668.

_____. 1982. "Pulp, Paper and Paperboard and the Builders' Paper and Board Mills Point Source Categories Effluent Limitations Guidelines, Pretreatment Standards and New Source Performance Standards." 47:52006-52070.

_____. 1985. "Intent to List Chloroform as a Hazardous Air Pollutant." 50:39626-39634.

_____. 1986(a). "Hazardous Waste Management System: Identification and Listing of Hazardous Waste; Notification Requirements; Reportable Quantity Adjustments; Proposed Rule." 40:21648-21693.

_____. 1986(b). "Pulp, Paper and Paperboard and the Builders' Paper and Board Mills Point Source Categories: Best Conventional Pollutant Effluent Limitations Guidelines." 51:45232-45242.

_____. 1986(c). "Review and Amendment of Standards of Performance for New Stationary Sources; Kraft Pulp Mills." 51:18538-18545.

_____. 1987. "Revisions to the National Ambient Quality Standards for Particulate Matter." 52:24634-24666.

_____. 1988. "Agency Information Collection Activities Under OMB Review (Study on Dioxin Formation During Bleaching of Wood Pulp)." 53:3937.

Fraas, Arthur G., and Vincent G. Munley. 1989. "Economic Objectives Within Bureaucratic Decision Process: Setting Pollution Control Requirements Under the Clean Water Act." *Journal of Environmental Economics and Management*, 17:35-53.

Fuller, B., et al. 1976. "Preliminary Scoring of Organic Air Pollutants." Report to the U.S. Environmental Protection Agency, Office of Air Quality Planning and Standards. EPA-450/3-77-008 a-e. Mitre Corporation, McLean, VA.

Jordan, Bruce C., Harvey M. Richmond, and Thomas McCurdy. 1983. "The Use of Scientific Information in Setting Ambient Standards." *Environmental Health Perspectives*, 52:223-240.

Magat, Wesley A., Alan J. Krupnick, and Winston Harrington. 1986. *Rules in the Making: A Statistical Analysis of Regulatory Agency Behavior.* Resources for the Future, Washington, DC.

Marcus, Alfred A. 1980. *Promise and Performance: Choosing and Implementing an Environmental Policy.* Greenwood Press, Westport, CT.

McGarity, Thomas O. 1983. "Media-Quality, Technology, and Cost-Benefit Balancing Strategies for Health and Environmental Regulations." *Law and Contemporary Problems*, 46:159-233.

Pahl, Dale. 1983. "EPA's Program for Establishing Standards of Performance for New Stationary Sources of Air Pollution." *Journal of the Air Pollution Control Association*, 33:468-482.

U.S. Department of Health, Education, and Welfare (HEW). 1969. "Air Quality Criteria for Particulate Matter." #AP-69. National Air Pollution Control Administration, Washington, DC.

U.S. Environmental Protection Agency (EPA). 1979. "Kraft Pulping: Control of TRS Emissions from Existing Sources." EPA-450/2-78-0036. Research Triangle Park, NC.

_____. 1980. "Dioxins." EPA-600/2-80-197. Washington, DC.

_____. 1987. "The National Dioxin Study. Tiers 3, 5, 6, and 7." EPA-440/1-87-003. Washington, DC.

_____. 1988. "U.S. EPA/Paper Industry Cooperative Dioxin Screening Study." EPA-440/1-88-025. Washington, DC.

Zaragoza. Lawrence J. 1982. "The Use of Biological Information in the Development of National Ambient Air Quality Standards." Ph.D. Thesis. University of California. Los Angeles, CA.

Chapter 3

TECHNOLOGY APPROACH

The Clean Water Act establishes the national goal of achieving fishable and swimmable water quality and mandates technology-based standards as the primary regulatory approach for reducing discharges. These rules apply uniformly without regard to their effects on local ambient water quality. The Act imposes uniform technology-based standards that do not account for differences in compliance costs, existing stream quality, contributions of other effluent sources, or recreation potential. Thus, we would expect that the relation between benefits and costs would vary widely across sites.

An assessment of the efficiency of a technology-based standard in reducing environmental risks requires linking a specific regulatory program, the costs of implementing it, and the benefits that accrue from it. The linkages for regulation of conventional pollutants from the pulp and paper industry are illustrated in Figure 3-1. This figure shows that, when implemented, a regulatory program based on technology-based standards will impose pollution control costs and reduce discharges from pulp and paper mills. These discharge reductions, taking into account hydrological, geological, and man-made features of a river (e.g., other industrial facilities or sewage treatment plants), may improve water quality in the receiving reaches. As water quality improves, ecological habitats become more suitable for recreation activities, especially fishing and swimming. The improved recreation potential will increase people's use of rivers. Monetary valuations of the benefits from water quality improvements are based on estimates of people's willingness to pay for the enhanced quality and increased frequency of these recreation activities.

The real resource costs required for major segments of the pulp and paper industry to comply with technology-based standards are those expenditures in excess of the credits for materials and energy recovery.[1] They are unadjusted for tax credits, depreciation allowances, and less than market rate loans, which makes them more than the "actual" (cash flow) costs to the industry.[2] The effluent reductions analyzed are biochemical oxygen demand (BOD) and total suspended solids (TSS). The costs and effluent reductions result from the compliance by existing mills with

Figure 3-1. Framework for Benefit-Cost Analysis of
the Clean Water Act Technology-Based Standards

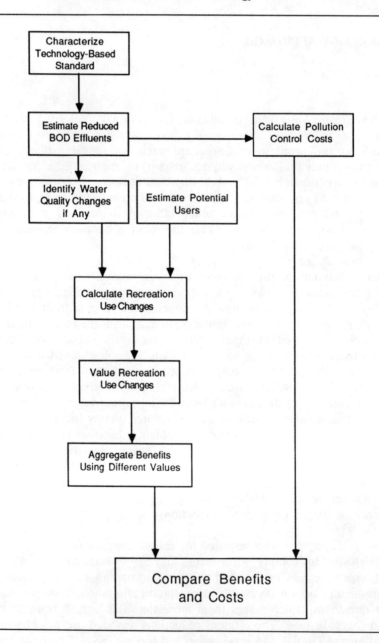

effluent limits specified in discharge permits and by new mills with effluent limits specified in New Source Performance Standards.

There are three categories of benefits resulting from reduced effluent loadings from pulp and paper mills: reduced human health risks from contaminated drinking water supplies, enhancement of water-based recreation, and ecological benefits. Ecological benefits are not currently monetizable and are therefore not included in this study. Nevertheless, these benefits may be significant in some cases and should be included in a full assessment of benefits and costs. The most obvious case for inclusion is the ecological benefits of the 30 percent reduction in chlorinated organic compounds associated with meeting current effluent limitations for conventional pollutants. To the extent that available data permit, we describe the potential benefits and costs of significant reductions in chlorinated organic compounds in Chapter 8. Because few pulp and paper mills are located upstream of drinking water intakes, the following analysis addresses only benefits associated with recreation activities.[3]

POLLUTION CONTROL COSTS AND EFFLUENTS

We estimate the real resource costs (in 1984 dollars) and effluents at three points in time -- 1973, 1984, and 1994.[4] Pollution control costs incurred before 1973 are not considered to be associated with the Clean Water Act because it was not passed until the end of 1972. Instead, they resulted from state and local agency requirements. Costs in 1984 are the current estimates of costs for water pollution control, resulting from actions before 1973 and actions in response to the 1972 Clean Water Act. Costs in 1994 are the estimated future costs of pollution control, taking into account increased production but assuming the same effluent limits as in 1984. Other types of costs and effluent estimates are those associated with bringing existing mills into compliance with their permit limits, those associated with 1984 production in the absence of federal regulations, and those associated with technology-forcing regulations by 1994.[5]

The main text of this chapter reports only the cost and effluent estimates for mills that discharge effluents directly to a water body. These data are central to the efficiency critique of the technology-based standard. We did estimate, however, the costs for indirect-discharging mills, those that discharge to a jointly owned or a municipal treatment system. We summarized our findings for these mills in Appendix 3-A.

Inventory -- 1973, 1984, and 1994

We identified 325 directly discharging pulp and paper mills operating in 1973 and 334 mills in 1984 and projected there would be 345 mills operating in 1994.[6] We could find sufficient data for characterizing pollution control costs and effluent discharge, however, for only 306 mills in 1973, 306 mills in 1984, and 318 mills in 1994.

We describe the procedures used for establishing this inventory in Appendix 3-B and list the mills covered in Appendix B. The background data for these mills are presented jointly for two of the three snapshot years, 1973 and 1984, and separately for 1994. The data categories for each year are mill type, mean daily production, pollution control equipment in place, and permit limits.

Effluent Data

For the three snapshot years -- 1973, 1984, and 1994 -- we drew on several sources to characterize effluents from directly-discharging mills.[7] For alternative regulatory conditions (full compliance with permits or technology forcing by 1994), we used the Cost of Clean model, which is described in the next section, to calculate effluent discharge.

Cost Modeling

We constructed the Cost of Clean model to estimate plant-specific costs and effluents for points in time and regulatory conditions [Putnam, Hayes, and Bartlett, Inc. 1986]. Figure 3-2 illustrates the basic structure of the model with its distinct parts -- baseline, permit compliance, and future regulation.

Baseline Costs. The first part of the model computes the capital, operation and maintenance, and annual costs of the effluent treatment processes that are in place in each of the three baseline years: 1973, 1984, and 1994. (See Appendix 3-C for a description of the cost functions.) Parallel to the computation of the treatment costs for each of the three baseline years, the model incorporates or calculates the effluent reductions and the resulting effluent levels and reports them in both pounds of conventional pollutants per ton of paper production and tons of pollutants in the given year.

Figure 3-2. Simplified Flowchart of the Cost of Clean Model -- Water

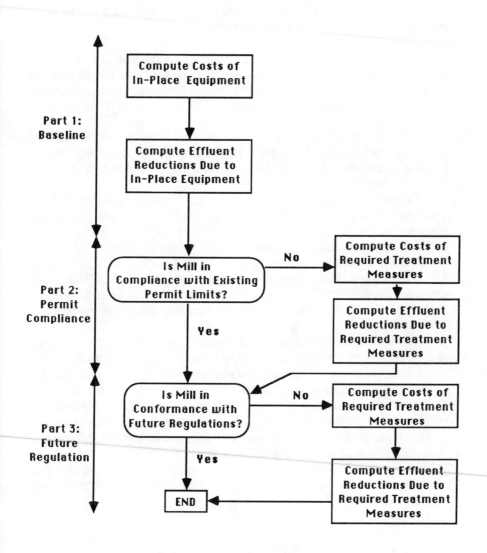

The costs of conventional pollution control for 1973 represent an estimate of the costs of equipment installed at mills for the purposes of effluent reductions. This equipment was installed as a result of state and local water pollution control regulations before any substantial requirements of the Clean Water Act.

The costs for 1984 represent an estimate of the costs of equipment in place in 1984. These costs resulted from standards associated with the Clean Water Act and with regulations imposed by state and local agencies in the absence of Federal requirements.

The costs of pollution controls for 1994 combine the 1984 costs with the additional costs of expanded and new mills meeting New Source Performance Standards. Costs and effluents are estimated accordingly.

Compliance Costs. The model also assesses the compliance costs for a mill not in compliance with its 1984 permit limits. If a mill's effluent levels are below the permit limitations, then the mill is considered to be in full compliance, and nothing more is done with the mill in this part of the model. If, however, one or both of the mill's effluent levels for BOD and TSS exceed their respective permit limitations, then the model estimates the additional treatment processes and associated costs required to bring the mill into compliance and the resulting reductions in effluents.[8]

Costs of More Stringent Regulations. The third part of the model begins by comparing the mill's effluent levels with the future effluent limitations of interest. Currently, these effluent limitations equal the Best Conventional Technology Option 1 levels analyzed in the Best Conventional Technology Development Document [EPA. 1986(b)]. If effluent levels for both BOD and TSS are below these levels, then a mill is considered to be in conformance and the mill is not further analyzed. If, however, one or both of the effluent levels exceed the allowable levels, then the mill is considered to be non-conforming. The model then calculates both the costs of installing the Best Conventional Technology Option 1 limitations and the resulting reductions in effluent levels.

Estimated Costs and Effluents

This section presents national summaries of the estimated water pollution control costs and effluents for 1973, 1984, and 1994 for most direct-discharging pulp and paper mills. Cost estimates are presented for capital costs, operation and maintenance costs, and annual costs. The annual costs are the sum of the annualized capital costs and operation and maintenance costs. The effluents are the actual or calculated discharges for each year under different conditions for the period 1973 to 1994.

Some care is needed in interpreting the cost results, particularly those attributable to the Clean Water Act. To address this issue, we first calculated the actual costs in 1973 and 1984. We then estimated costs (and effluents) without the Clean Water Act by combining 1984 production levels with 1973 treatment processes. The without-Clean Water Act costs were then subtracted from the with-Clean Water Act costs, where costs represent 1984 production with 1984 pollution control. This procedure eliminates those treatment costs already incurred before 1973, plus any expansion in treatment costs due to increased production. The costs attributable to the Clean Water Act in the 1973-1984 period then reflect only the real resource costs for the industry to meet Best Practicable Technology standards.

1973. The annual pollution reduction costs and discharges of conventional pollutants by the pulp and paper industry in 1973 are listed in Table 3-1. These costs were approximately $580 million and were the result of state and local actions. The costs attributable to direct dischargers accounted for approximately 90 percent of the combined costs of direct and indirect dischargers. The industry discharged an estimated 700,000 tons of BOD and 680,000 tons of TSS.

1984 without the Clean Water Act. The annual costs in 1984 if the Clean Water Act requirements had not been implemented (but with state and local requirements) would have been about $700 million. The 1984 water effluents without the Clean Water Act would have been 1,100,00 tons of BOD (1.6 times 1973 effluents) and 1,090,000 tons of TSS (1.7 times 1973 effluents).

1984 with the Clean Water Act. The annual costs in 1984 with the Clean Water Act requirements were approximately $1,010 million compared to an estimated $700 million without the Clean Water Act. This additional $310 million investment in pollution controls resulted in a 500,000-ton decrease in BOD and a 490,000-ton decrease in TSS instead of a 400,000-ton increase in BOD and a 410,000-ton increase in TSS between 1973 and 1984.

Table 3-1. National Summary of Water Pollution Control Costs and
Effluents for Direct-Discharging Mills in This Study ($1984)

	Costs (10^6)[a]			Effluents (tons 10^3)	
Scenario	Capital	O&M	Annual	BOD	TSS
1973[b]	$2,620	$235	$580	700	680
1984 w/o CWA	$3,225	$280	$700	1,100	1,090
1984 w/ CWA[b]	$4,490	$420	$1,010	200	190
1984 w/ full compliance	$4,565	$430	$1,030	180	180
1994	$5,150	$490	$1,170	230	230
1994 w/ BCT	$5,530	$540	$1,270	170	160

a. Annual costs are the sum of annualized capital costs [capital times a capital
recovery factor of 0.1315 (which represents a 10 percent real rate of interest
and a 15-year depreciation period)] and operation and maintenance costs.
b. The pollution control cost per ton of product in 1973 was $15. It increased
to $22 by 1984.

1984 with Compliance. The annual costs in 1984 if full compliance
with the Clean Water Act had been attained would have been $1,030
million. The incremental annual costs of $20 million would result from
correcting violations of national permit limits at 49 mills. The incremental
reduction in BOD would have been 20,000 tons, and the incremental
reduction in TSS would have been 10,000 tons.

1994. The annual costs in 1994 to meet existing and new mill
effluent limits are projected to be approximately $1,170 million. The
annual costs would be 1.2 times the 1984 annual costs. In spite of these
additional expenditures, total effluents would increase as a result of a 15
percent increase in capacity. The discharges of BOD and TSS are
projected to increase by 15 and 20 percent, respectively.

1994 with Best Conventional Technology. The annual costs with a
minimal technology forcing requirement, defined here as the Best
Conventional Technology Option 1, are projected to be about $1,270

million. The incremental annual costs of $100 million would result in a 60,000-ton reduction of BOD and a 70,000-ton reduction of TSS in the 1994 baseline estimates. The Best Conventional Technology limits would reduce effluents at 176 mills.

The cost and effluent estimates do not represent the total costs of pollution control or total effluent reductions on the part of the entire industry. However, they represent most of the costs and effluents of direct-discharging mills based on comparisons with total industry production in 1973 and 1984. The mills included in this study accounted for 85 percent of the 1973 U.S. paper and paperboard production and 83 percent of the 1984 U.S. paper and paperboard production [Medynski. 1973, Brandes. 1984].

Cost Comparisons

We compared our real resource cost estimates with other real resource and financial cost estimates. Other real cost estimates are in "The Cost of Clean Air and Water Report to Congress 1984" [EPA. 1984] and the "Development Document for Best Conventional Pollutant Control Technology Effluent Limitations Guidelines for the Pulp, Paper, and Paperboard and Builders' Paper and Board Mills Point Source Categories" [EPA. 1986(b)]. The only credible estimate of financial costs is the "Pollution Abatement Costs and Expenditures, 1984" [U.S. Bureau of the Census. 1986]. We present the results of the comparison in Appendix 3-D. In general, the real resource cost estimates from the Cost of Clean model are similar to the 1984 EPA cost estimates and are about 25 percent less than the engineering cost estimates in the 1986 EPA Development Document. The Cost of Clean model real resource cost estimates are approximately double the financial cost estimates reported by the Bureau of the Census. Real resource cost estimates should be higher than financial cost estimates because the latter reflects the effects of federal and state taxes, accelerated depreciation provisions, industrial revenue bonds, and subsidies via the municipal wastewater treatment program.

WATER QUALITY IMPACTS

The benefits of improved water quality under the Clean Water Act must be measured by changes in specific water quality parameters. In general, such parameters should be relevant to the actual effluent discharged and should influence the potential uses of the stream. For this study, dissolved oxygen was chosen as the indicator variable for overall water quality. Dissolved oxygen is a water quality parameter directly related to relevant stream uses, and carbonaceous BOD is the primary effluent that affects the in-stream dissolved oxygen. Although other water quality parameters also vary and can influence possible stream uses, lack of available data and the complexity of physical processes made their inclusion infeasible for a nationwide study.

Sample Sites

The difficulty of assessing the water quality effects of effluent loadings from one source type in river systems with several effluent source types necessitated our limiting the benefit assessment to rivers where pulp and paper mills are the dominant point source of pollution. First, sites were chosen where no other industrial or municipal dischargers were located within 10 miles upstream or downstream of a mill. This resulted in a sample of only 45 mills on 45 reaches. Mills were then added if the only dischargers within the 10-mile distance were municipal wastewater treatment plants because information on the discharge characteristics of these plants is generally readily available. This increased the sample size to 84 mills on 82 reaches. Some of these mills were eliminated, however, due to the difficulty in modeling discharges to lakes and oceans. This resulted in 68 "sample mills" on 66 reaches that were selected for water quality modeling. The geographic distribution of the 306 mills that directly discharge and the 68 mills that were selected for modeling is shown in Figure 3-3.

As stated earlier, effluent loadings from 306 directly discharging pulp and paper mills were modeled for the year 1984. To determine whether the assimilative capacity of the receiving waters for our sample of 68 mills was representative of the assimilative capacity of the receiving water for all mills in the industry, the distribution of dilution ratios was compared for both sets of mills. The dilution ratio is a measure of the relative mean flow from a discharger to the flow in the receiving stream and is calculated as the ratio of stream flow to discharge flow. This value was calculated for each sample mill and for all mills that discharge into free-flowing streams (285 out of the 306 mills).

The resulting histogram of dilution ratios is presented in Figure 3-4. As illustrated, there is some difference between the sample and overall population of mills discharging into free-flowing streams. Although in

Figure 3-3. Map of Pulp and Paper Mills Included in the Water Analysis

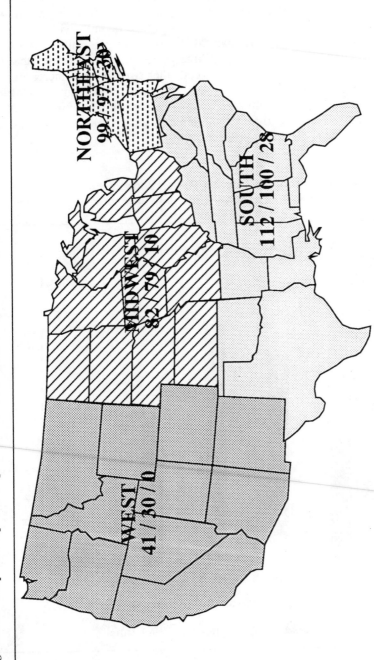

NORTHEAST
99 / 97 / 30

MIDWEST
82 / 79 / 10

SOUTH
112 / 100 / 28

WEST
41 / 30 / 0

LEGEND: No. of direct discharging mills / no. of mills with cost estimates / no. of mills in benefit-cost analysis

both cases approximately 60 percent of the mills have a dilution ratio between 10 and 1,000, a larger percentage of the sample mills has a lower dilution ratio (<10) than the overall population of mills. Similarly, the overall population contains a greater percentage of mills with a dilution ratio greater than 1,000 than does the sample mill population (approximately ten percent versus three percent). This indicates that the sample mills discharge, on average, to smaller streams, thus increasing the likelihood that a change in water quality would result from a change in their effluent loadings. For this reason, average benefits for this sample could tend to be larger than for the industry as a whole. However, other factors may work in the opposite direction.

Figure 3-4. Comparison of Dilution Ratios Between
              Sample Mills and All Mills (low flow scenario)

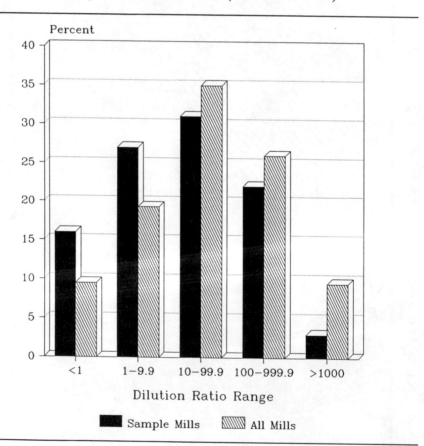

Scenarios

Once the mills were selected, we estimated BOD effluents for three scenarios. Scenario 1 portrays 1973 actual effluents from pulp and paper mills and municipal wastewater treatment plants where appropriate. It approximates conditions before the Clean Water Act by applying 1973 effluent limitations to 1973 pulp and paper mill production and 1973 municipal population served. Scenario 2 portrays 1984 hypothetical effluents from pulp and paper mills and municipal wastewater treatment plants. It approximates effluent discharge without the Clean Water Act by applying 1973 effluent limitations to 1984 pulp and paper mill production and 1984 municipal population served. Scenario 3 portrays 1984 actual effluents from pulp and paper mills in compliance with their permit limits and municipal wastewater treatment plants based on Discharge Monitoring Reports filed with EPA. It approximates effluent discharge with the Clean Water Act's effluent limitations applied to 1984 pulp and paper mill production and to 1984 municipal population served. Effluent loadings for the 68 pulp and paper mills and the relevant municipal wastewater treatment plants for the three scenarios are presented in Appendix 3-E.

Water Quality Modeling

The BOD-dissolved oxygen relationship is a well-accepted and well-documented process that has been in common use for over half a century. Organic waste (BOD) in the effluent exerts an oxygen demand on the receiving stream. This reduces the amount of oxygen available to sustain aquatic plant and animal life [EPA. 1986(a)]. Natural physical and biological processes replenish the oxygen in the stream so that an oxygen sag and subsequent recovery are experienced downstream from an effluent.

The water quality model used in this study is a modified version of the routing component of EPA's Routing and Graphical Display System (RGDS) [Grayman. 1985]. This model uses the Streeter-Phelps formulation to predict the interaction of dissolved oxygen and BOD in a free-flowing stream. The low flow (the minimum value for the seven day moving-average over a 10 year period) and average summer flow stream conditions were used to model these reaches. Transport and transformation characteristics of the stream are additional inputs to the model. The model and data inputs are briefly described in Appendix 3-F.

Figure 3-5 illustrates the BOD-dissolved oxygen relationship. The addition of oxygen-demanding material is responsible for the magnitude

Figure 3-5.   Water Quality Impacts on Stream Usage

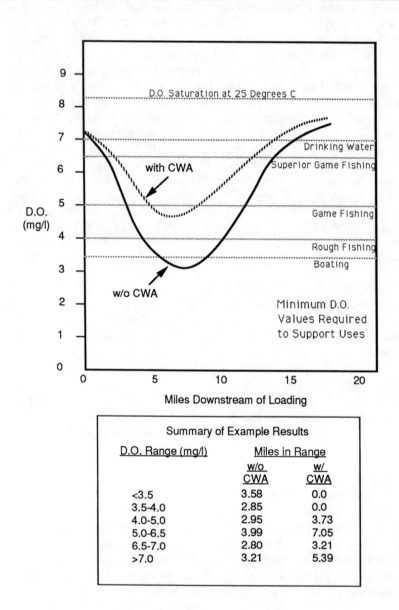

of the drop in dissolved oxygen downstream from the source. The natural assimilative capacity of the stream is responsible for the recovery of in-stream dissolved oxygen. If the amount of BOD discharged is reduced, then on average the magnitude of the drop in dissolved oxygen is reduced, and the overall level of dissolved oxygen in the affected stream reach increases.

As shown in Figure 3-5, various levels of dissolved oxygen will sustain different activities. A stream with a sustained level of dissolved oxygen of 5.0 mg/l will support game fishing. Lesser quality is required to support boating, whereas higher quality is desirable for superior game fishing. An overall water quality index, a modified version of the Water Quality Index of Mitchell and Carson [1981], is used to summarize the results for each reach. The feasible activity level for the overall reach is based on the following classification:

Dissolved Oxygen Range	Activity Levels
< 3.5 mg/l	unusable
3.5 - 4.0 mg/l	boatable
4.0 - 5.0 mg/l	rough fishing
5.0 - 6.5 mg/l	game fishing & swimming
> 6.5 mg/l	superior game fishing & swimming

The lowest dissolved oxygen segment with affected stream miles determines the activity classification for the reach. In the case illustrated in Figure 3-5, the reach would be classified as unusable without the Clean Water Act and suitable for rough fishing with the Clean Water Act.

The profiles of four of the 66 segments are presented in Table 3-2 to show the results of the model and how we interpreted them for benefit-cost analysis. For example, effluent from Mill 343 affected water quality in 45.1 downstream miles. The 1973 effluent significantly impaired downstream water quality, with 29.6 miles showing a dissolved oxygen value less than 3.5 mg/l and resulting in a classification as unusable. The hypothetical 1984 effluent would have more severely impaired downstream water quality than the 1973 effluent because production increased by 50 percent between 1973 and 1984. The stream would be classified as unusable. The actual 1984 effluent, a 90 percent reduction in loadings, eliminated all stream impairment below 6.5 mg/l. The stream would be classified as suitable for superior game fishing. Similar interpretations characterize the results at the other three mills.[9]

Table 3-2. Impacts of Three Scenarios for Four Mills on Water Quality and Activity Levels

| Scenario | Total Miles | Miles of Streams with Dissolved Oxygen Range | | | | | | Activity Levels |
		<3.5	3.5-3.9	4.0-4.9	5.0-6.4	6.5-7.0	>7.0	
Mill #22								
1973	77.7	0.0	0.0	41.8	29.8	6.1	0.0	rough
1984 w/o CWA	77.7	40.0	10.0	10.0	14.1	3.6	0.0	unusable
1984 w/ CWA	77.7	0.0	0.0	0.0	57.5	20.2	0.0	game
Mill #161								
1973	50.2	0.0	0.0	0.0	41.1	9.1	0.0	game
1984 w/o CWA	50.2	0.0	0.0	0.0	50.2	0.0	0.0	game
1984 w/ CWA	50.2	0.0	0.0	0.0	25.7	24.5	0.0	game
Mill #216								
1973	64.7	29.6	2.9	4.3	8.6	9.2	10.1	unusable
1984 w/o CWA	64.7	31.1	2.7	3.9	8.5	8.9	9.6	unusable
1984 w/ CWA	64.7	17.6	4.0	5.7	11.4	10.0	16.0	unusable
Mill #343								
1973	45.1	29.6	6.1	6.5	2.1	0.7	0.1	unusable
1984 w/o CWA	45.1	42.3	0.4	0.8	1.2	0.3	0.1	unusable
1984 w/ CWA	45.1	0.0	0.0	0.0	0.0	7.9	37.2	s. game[b]

a. Milligrams per litre.
b. Superior game fishing.

Estimated Water Quality Improvements

The summary results of the water quality simulations without (Scenario 2) and with (Scenario 3) the Clean Water Act effluent limitations for 68 mills on 66 river segments show varied outcomes (Figure 3-6). Twenty-five mills on 25 segments (about 35 percent of the total number of mills) discharge into rivers that remain in the less than superior game fishing category. Of these, 15 mills on 15 segments (about 20 percent of the total number of mills) discharge into rivers that remain in the unusable category under both scenarios. One mill on one segment discharges into a river that remains in the rough fishing category under both scenarios. Remaining in the unusable or rough fishing use categories under the 1984 with the Clean Water Act scenario indicates that current

permit limits on discharge are inadequate and that additional effluent
reductions are needed to achieve water quality that would support game
fishing recreation. Nine mills on nine segments remain in the game fishing
category under both scenarios. Falling in the game fishing use category
under the 1984 with Clean Water Act scenario suggests that the permit
limits should be reviewed to determine if additional effluent reductions are
needed to enhance game fishing potential.

Figure 3-6. Potential Recreation Activities With and Without the
Clean Water Act

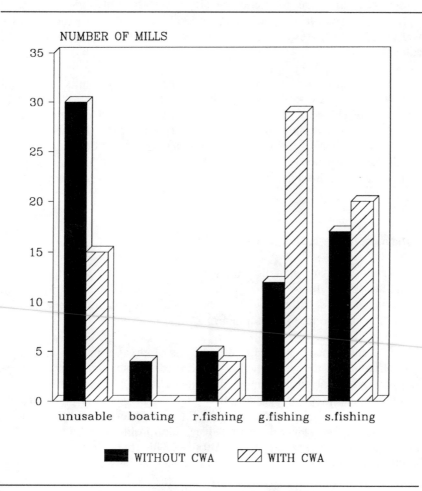

Seventeen mills on 15 segments (25 percent of the total number of mills) discharge into rivers that fall into the superior game fishing category under both scenarios. Falling into the superior game fishing use category under the 1984 without Clean Water Act scenario, which represents the highest loadings among the three scenarios, indicates that the current discharge restrictions were unnecessary to maintain satisfactory water quality.

Twenty-six mills on 26 segments (almost 40 percent of the total number of mills) discharge into rivers that change use categories under the two scenarios. Moving to a higher use category indicates that the Clean Water Act discharge restrictions were necessary, although not always sufficient, to achieve satisfactory water quality. At three of the 26 mills, additional effluent restrictions are needed to reach the game fishing use category.

The primary factors determining the water quality results are the in-stream capacity to assimilate organic waste and the reduction in mill discharge of effluents. In general, the mills whose water quality situation is classified as unusable under both scenarios discharge into rivers with limited assimilate capacity and did not significantly reduce their waste loadings. Mills whose water quality situation is classified as superior game fishing under both scenarios discharge into rivers with greater assimilative capacity.[10]

Validation of the RGDS Model

Inherent in any modeling applications are necessary assumptions and simplifications. In order to develop confidence in the validity of the modeling effort and the results of the model's application, a method for validating the model is good practice.

The model used to predict the impacts of changes in pulp and paper mill effluents on the water quality of the receiving stream, as measured by the level of dissolved oxygen, used standard relationships that have been commonly accepted for over a half century. Due to the large geographic scope of the study, however, many of the parameters used by the model were based on general nationwide relationships, rather than on field-checked, site-specific information. This data limitation further emphasizes the need for validating the model.

For validation, we compared the model's predicted values with those from a large water quality data base [Grayman. 1987]. Details of this assessment are presented in Appendix 3-G. The analysis was in general very supportive of the modeling results. Namely, in most cases where a mill was identified with modeled improvements in dissolved oxygen from 1973 to 1984 and field data were available, the field data supported this improvement. Similarly, for mills with no modeled improvement in dissolved oxygen from 1973 to 1984, the field data indicated either decreases in dissolved oxygen or only modest increases in dissolved oxygen. Certain limitations are inherent in the analysis because of the limited amount of data. At a minimum, however, the analysis of the available field data does not contradict the modeling results and more likely strongly supports the results.

MONETARY BENEFITS

Society values improvements in water quality because the improvements restore recreation potential and ecological productivity. Estimating societal values in monetary terms for improvements in recreation potential requires calculating the size of the potential recreation market and measuring the willingness-to-pay of recreators for the services of a given stream [Smith and Desvousges. 1986].

Of the 68 segments for which water quality changes were simulated, 26 indicate significant improvements in water quality. Benefits for the other 42 are estimated to be zero or minimal because the potential recreation use category did not change as a result of the Clean Water Act effluent restrictions. The 68 segments differ in significant ways from rivers for which monetary benefits have been estimated. Consequently, we were forced to transfer benefit estimates from existing studies to each of the 68 reaches.

Recreation Population

The potential recreation population that would benefit from water quality improvements at a given site depends upon the size of the population in the vicinity of the site, the attractiveness of the site, and availability of comparable substitute sites. We attempted to obtain site-specific information on these factors by reviewing State Comprehensive Outdoor Recreation Plans and surveying state and local recreation planners. We could find quantitative estimates of recreation visitation at only four of the six sites of regional significance.[11] For the remaining two sites of regional significance and the 60 sites of only local significance, we developed a general method to estimate visitation.

We estimated recreation visitation at these remaining sites by first measuring the size of the potential recreation market and then by calculating visitation based on an index of availability of potential recreation sites. The size of the recreation market is the number of households living within a 30-mile radius of a site.[12] We assumed that all households in the market participated in river-based recreation.

The visitation at a given site depends on its attractiveness compared to other sites and the participation rate of the general population in water-based recreation. The attractiveness of a given site depends on the quantity and quality of available substitutes for the site. We obtained a crude index of available substitutes for each site by calculating the ratio of total stream miles within a 30-mile radius to the stream miles affected by discharge from the mill. The distribution of index values roughly clustered in three ranges, which were used to define few, moderate, and many substitute categories.

As a first approximation of visitation, we assumed 100 percent participation in river-based recreation and that each site with a moderate number of substitutes would receive visits from a maximum upper bound of 30 percent of households living within 30 miles. Sites with few substitutes were assigned a maximum visitation rate of 50 percent, and sites with many substitutes were assigned a maximum visitation rate of 10 percent. The results of this assumptions are presented in Appendix 3-H. These assumptions overstate likely visits by a wide margin in order to ensure that our benefit estimates constitute an upper bound. The sensitivity of our conclusions to these assumptions is discussed below.

Valuation of Recreation Benefits

Although there are no specific studies of recreation benefits for the river segments under consideration, we could extrapolate from studies of other rivers. The extrapolations generally follow procedures outlined in Freeman [1984], Naughton and Desvousges [1986], and Naughton, Desvousges, and Parsons [1987]. We deviate in some respects where we have better data. The findings of the extrapolations are reported in Appendix 3-I.

Even the short list of existing studies does not provide an ideal basis for extrapolating values for improvements in water quality on the 68 reaches.[13] Therefore, we used plausible lower- and upper-bound values in view of the limitations of the existing studies and extrapolation procedures. Table 3-3 shows the willingness-to-pay values per household per year used for the 68 reaches. Because contingent value estimates include both use and non-use values, we took the travel cost estimates as the lower bound and the contingent valuation estimates as the upper bound of the plausible range of benefits. We interpolated from existing studies to obtain values for each possible water quality change. Differences in site characteristics introduce greater uncertainty which is reflected in the wider range of values assumed for lower value categories of improvement. The lower and upper bounds reflect plausible bounds for recreation values, given the range of site characteristics found at the sites affected by the 68 mills.[14]

As can be seen in Table 3-3, we assigned values to intra-use as well as inter-use category changes. Intra-use category values reflect a change

Table 3-3. Assumed Transfer Values for Water Quality Benefits (Willingness-To-Pay Per Household Per Year) ($1984)

Initial Water Quality	Final Water Quality	Willingness-to-Pay	
		Lower Bound	Upper Bound
U	U	$1-3	$9-18
U	B	$5	$35
U	R	$15	$50
U	G	$20	$80
U	G*	$25	$90
B	B	$2-4	$8-15
B	R	$8	$30
B	G	$15	$50
B	G*	$20	$60
R	R	$3-5	$6-13
R	G	$10	$25
R	G*	$15	$35
G	G	$3-6	$5-10
G	G*	$12	$20

U = unusable   B = boatable   R = rough fishing
G = game fishing   G* = superior game fishing

in water quality, but one that was not sufficient to change potential use categories. Decreases in effluents attributable to the Clean Water Act at 25 mills on 25 segments reduced the affected stream mileage in the assigned use category and increased the stream mileage in more beneficial use categories, but were not sufficient to change the use category. If the stream mileage changed by less than 25 percent, we arbitrarily assigned 25 percent of the value in the next highest use category change; if the stream mileage changed by more than 25 percent, we arbitrarily assigned 50 percent of the value in the next highest use category. For example, we assigned lower bound values of $1 to $3 and upper bound values of $9 to $18 to an U-U change. We did not assign any value to intra-use category movements in the superior game fishing category (G*) because we could not imagine any measurable improvement in ecological activity, and thus recreation activity, with dissolved oxygen changes above 6.5 mg/l.

Estimated Benefits

Table 3-4 combines benefits per household with the number of households to obtain the total willingness to pay for water quality improvements and compares recreation benefits to pollution reduction costs. The lower bound benefit estimates reflect only use values and the upper bound benefit estimates reflect both use and non-use values. The cost estimates are the differences in costs between the with and without Clean Water Act scenarios. Our analysis indicates that 11 mills on rivers with changes in potential recreation activities have estimated upper-bound benefits greater than the pollution control costs attributable to the Clean Water Act. However, only seven of these mills clearly satisfy the benefit-cost criterion, with costs that are less than the lower-bound benefits. The remainder of the 68 sites have costs that exceed the upper-bound benefits.

Appendix 3-J shows the results of sensitivity tests of our treatment of intra-category changes. If all intra-category changes are valued at the upper-bound willingness to pay to move to the next higher use, benefits rise only from $56.7 to $62.9 million.

Possible Biases in the Benefit Calculations

There are several sources of possible bias in our reported benefit calculations. These include use of low-flow simulations, unrepresentativeness of the sample reaches, inconsistencies between technical water

Table 3-4. Estimated Benefits and Costs of the Clean Water Act for 68 Pulp and Paper Mills ($1984 10^3)

Mill	Water Quality	Weights	Annual Benefits Min[a]	Annual Benefits Max[a]	Annual Costs
14	G-G	0.25	$10	$20	$220
16	G-G	0.25	$20	$30	$600
21	U-U	0.25	$10	$60	$940
22	U-G	1.0	$3,440	$13,760	$1,550
39	G-G	0.25	$20	$30	$400
42	U-U	0.25	$20	$130	$420
43	U-U	0.25	$0	$20	$500
44	U-U	0.25	$10	$70	$100
48	U-G	1.0	$150	$610	$7,200
51	U-R	1.0	$30	$100	$600
56	U-U	0.5	$70	$460	$500
57	G-G	0.25	$20	$30	$1,400
66	R-G	1.0	$1,380	$3,450	$200
77	U-U	0.25	$0	$30	$600
79	U-U	0.25	$10	$70	$100
82	U-U	0.25	$20	$150	$4,500
85	U-U	0.25	$20	$160	$2,200
87	G-G	0.50	$90	$140	$360
89	G-G	0.25	$20	$30	$3,500
91	B-G	1.0	$50	$180	$2,000
92	G*-G*	0.0	$0	$0	$990
97	G*-G*	0.0	$0	$0	$260
100	U-G	1.0	$640	$2,560	$3,800
102	R-G	1.0	$320	$800	$6,500
103	G-G*	1.0	$310	$520	$3,360
105	U-G	1.0	$40	$170	$3,200
106	G*-G*	0.0	$0	$0	$220
125	G*-G*	0.0	$0	$0	$70
130	U-G	1.0	$1,950	$7,800	$100
141	G*-G*	0.0	$0	$0	$230
147	U-G	1.0	$50	$210	$1,700
150	G*-G*	0.0	$0	$0	$1,100
151	G*-G*	0.0	$0	$0	$260
157	G-G	0.5	$30	$50	$150
159	R-G	1.0	$170	$420	$1,100
165	G-G*	1.0	$190	$310	$4,200
167	U-G	1.0	$60	$240	$4,100
169	G-G*	1.0	$1,100	$1,800	$400
170	G*-G*	0.0	$0	$0	$100

(cont.)

Table 3-4. (cont.) Estimated Benefits and Costs of the Clean Water Act for 68 Pulp and Paper Mills ($1984 10^3)

Mill	Water Quality	Weights	Annual Benefits Min[a]	Annual Benefits Max[a]	Annual Costs
182	U-R	1.0	$1,460	$4,870	$50
187	G*-G*	0.0	$0	$0	$170
192	U-U	0.5	$10	$100	$90
193	G*-G*	0.0	$0	$0	$40
194	B-G	1.0	$1,460	$4,870	$500
198	U-R	1.0	$100	$350	$900
201	G-G	0.25	$70	$120	$300
202	G*-G*	0.0	$0	$0	$40
206	G*-G*	0.0	$0	$0	$150
207	G*-G*	0.0	$0	$0	$400
209	G*-G*	0.0	$0	$0	$20
216	U-U	0.5	$60	$450	$1,400
217	R-R	0.25	$0	$10	$300
219	B-G	1.0	$990	$3,310	$800
235	U-G	1.0	$690	$2,770	$2,600
238	U-G	1.0	$230	$930	$3,000
260	U-U	0.25	$60	$450	$500
280	U-U	0.5	$10	$100	$150
286	G*-G*	0.0	$0	$0	$200
287	G*-G*	0.0	$0	$0	$10
288	B-G	1.0	$70	$220	$600
289	U-G	1.0	$300	$1,180	$900
293	U-U	0.25	$10	$100	$3,900
295	G-G	0.25	$10	$20	$250
296	U-U	0.5	$40	$290	$5,100
324	G*-G*	0.0	$0	$0	$630
343	U-G	1.0	$190	$770	$3,700
344	U-G	1.0	$170	$680	$650
355	R-G	1.0	$280	$710	$9,500
TOTAL			$16,510	$56,740	$96,570

a. The minimum value reflects use value only and the maximum value reflects use and non-use value.

Abbreviations:

         U = Unusable.
         B = Boatable.
         R = Rough fishing.
         G = Game fishing.
         G* = Superior game fishing.

quality parameters and user perceptions, and exclusion of possible nonuse benefits of nonusers. We have already noted that calculated benefits reflect low-flow water quality simulations rather than average-flow conditions. This choice is appropriate for some stream uses, especially fishing, but it tends to overstate improvements for an average year and biases benefits upward to some degree.

The number of households located within a 30-mile radius for the 68 mills is generally lower than for the 285 mills directly discharging into free-flowing streams as seen in Figure 3-7. On the one hand, 40 percent

Figure 3-7. Distribution of Households Around Mills

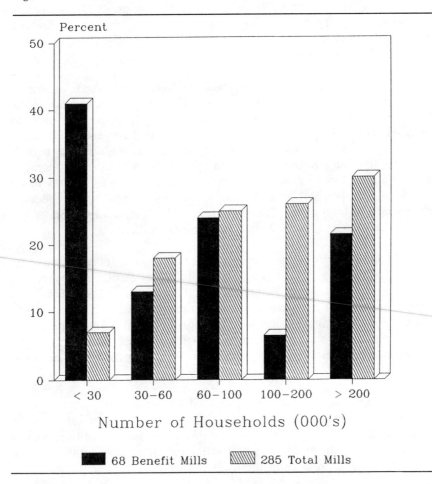

of the sample mills have fewer than 30,000 households located within this distance, whereas only 10 percent of the 285 mills have fewer than 30,000 households located within this distance. On the other hand, only 20 percent of the sample mills have more than 100,000 households located within this distance, but 50 percent of the 285 mills have more than 100,000 households located within this distance.

Because the sample mills are associated with smaller recreation markets, a given water quality improvement necessarily yields a lower monetary benefit. However, by assuming 100 percent participation rates, we have deliberately offset the effect of smaller populations by a considerable margin. Two national surveys found that between 50 and 60 percent of the population participates in swimming, fishing, or boating in oceans, lakes, and rivers [U.S. Department of Interior. 1979, President's Commission. 1987]. Approximately one-third of the households surveyed in the Monongahela study had recreated at one of 29 sites on the river [Smith and Desvousges. 1986].

We have also deliberately overstated visitation rates. An unpublished survey of Wisconsin lakes found that Lake Winnebago, which is the largest and most attractive lake in the state, received only 10 percent of total visits [Parsons. 1990]. In contrast, we have assumed 10 percent visitation rates for the least attractive sites in our sample. Since the benefit estimates are roughly proportional to the visitation rate assumptions, it would be necessary to raise these already overstated visitation rates by an additional 50 percent in order to make total upper-bound benefits comparable to total costs. This number of visits is clearly exaggerated for any of the streams affected by pulp and paper mills.

Note that imputing the contingent valuation estimates only to visitors excludes nonuse benefits for people who value water quality improvements, but do not visit the site. Although such benefits can be significant for sites of regional or national importance, nearly all the sites in our study are of only local interest. In order to raise upper-bound benefits to the level of costs, we would have to impute upper-bound willingness-to-pay values to a nonuser population somewhat more than half the size of the user population.

Although nonusers may assign little value to water quality improvements for any particular local site, they may enjoy nonuse benefits from the knowledge that national water quality in general has improved as a result of reduced pulp and paper effluents. If such benefits exist, we have been unable to account for them in this analysis.[15]

Another potential source of bias involves the correspondence between technical water quality parameters and the perceptions of users. We assigned broad water quality ranges to a small number of activity categories in order to use the few available estimates of water quality benefits. Again, we employed very liberal assumptions about user sensitivity to relatively small changes to ensure that we erred on the side of overstating benefits.

In the absence of reliable empirical data on visitor responses to water quality changes, we have had to employ various assumptions about water recreation participation rates, individual site visitation rates, and household willingness to pay for water quality improvements. We have taken care to ensure that the resulting benefit estimates overstate actual values by a substantial margin. Therefore, our subsequent conclusions are based on upper-bound estimates. For any given site that fails the benefit-cost criterion because its cost exceeds stated benefits, more accurate benefit estimates would not change our conclusion. If stated benefits exceed costs, we must be more tentative in drawing conclusions about the net benefits of the Clean Water Act.

FINDINGS

The Clean Water Act technology-based standards for the pulp and paper industry did not meet the benefit-cost criterion based on the 68 mills in our sample. As shown in Table 3-5, even the maximum benefit estimate, $56.7 million, is only 60 percent of the cost estimate. In addition, assuming the maximum benefit estimate, there were net benefits at only 11 out of the 68 mills. At 40 mills there were water quality benefits, but the net benefits were negative. And, at 17 mills, there were no discernable water quality benefits.

Earlier, we estimated the total costs attributable to the Clean Water Act's technology-based standards for controlling conventional water pollution from the pulp and paper industry at $310 million. To the extent that the 68 sample mills are representative of the 306 mills, the benefits of pollution control attributable to the Clean Water Act, assuming a maximum benefit estimate is most appropriate, are approximately 60 percent of the associated costs of pollution control.[16]

Table 3-5. Benefits and Costs Attributable to the Clean Water Act
Technology-Based Standard ($1984 10^6)

Valuation Scheme	Mills with B>C / Mills Analyzed	Total Benefits	Total Costs	Net Benefits
Minimum	7 / 68	$16.5	$96.6	($80.1)
Mid-Point [a]	7 / 68	$36.6	$96.6	($60.0)
Maximum	11 / 68	$56.7	$96.6	($39.9)

a. The mid-point value is the average between the minimum and maximum values.

This conclusion should not be surprising. The Clean Water Act requires the uniform application of technology standards as the minimum level of pollution control regardless of the environmental situation. One would expect some unnecessary pollution control. Nevertheless, the extent of over-control is surprising. Approximately 25 percent of the mills in the sample, and presumably 25 percent of all 306 mills, invested in pollution control that did not produce a significant environmental result. Also, one might be surprised at the extent to which the technology standards coupled with encouragement to require additional pollution control failed to achieve "fishable and swimmable waters" as mandated by the Clean Water Act. Approximately 25 percent of the mills in the sample, and presumably 25 percent of all 306 mills, would have to invest more in pollution control to eliminate serious water quality problems. Depending on the size of the recreation market, the availability of substitute recreation opportunities, and the size of the water quality change, some of these investments would be justified on benefit-cost grounds.

Notes

1. The EPA effluent guidelines for the pulp, paper, and paperboard industry cover four Standard Industrial Classification (SIC) codes: 2611 - pulp mills; 2621 - paper mills; 2631 - paperboard mills; and 2661 - building paper and roofing felt mills.

2. Hazilla and Kopp [1986] would argue that these real resource cost expenditures underestimate the true social costs of pollution control. They compared pollution control costs of the expenditure based and general equilibrium, "willingness to be compensated" approaches. They found that the expenditures approach compared to the general equilibrium approach overstates the true social costs of regulation in the early years and understates them in the out years. We did not attempt to estimate the true social costs because of the preliminary nature of their approach and of the difficulty of undertaking such an estimate for one sector of the economy.

3. Reduced health risks from toxic pollutants are not considered because only 12 out of 285 pulp and paper mills on free-flowing streams are located within 10 upstream miles of a drinking water intake, and none of those mills is among the 68 mills that are the focus of this benefit assessment. Moreover, control of toxic pollutants has yet to impose a significant cost on the pulp and paper industry.

The records of the 15 drinking water intakes downstream of the 12 pulp and paper mills did not reveal any concentrations of the toxic pollutants regulated by pulp and paper BAT regulations - pentachlorophenol, tricholorophenol, and zinc. The records also did not show, nor did we anticipate, concentrations of other total organo-chlorines, such as dioxin or chloroform. The downstream concentration of dioxin from a pulp and paper mill would be in the parts-per-trillion range and would not be detected by monitoring at drinking water intakes. The potential fish-consumption health risk, however, will be discussed in the chapter on toxic pollutants. The in-stream concentration of chloroform would be negligible because most of it volatilizes before it is discharged from a mill.

4. All costs in the report are in 1984 dollars. The Engineering News Record Cost index was used to change costs in dollars from any other year to costs in 1984 dollars [Appendix C].

5. This chapter does not estimate the costs or effluent reductions of toxics required to comply with the Best Available Technology effluent guidelines. EPA estimated that these guidelines would not require any additional pollution control expenditures for mills existing in 1984 and only minor costs compared to conventional pollution control costs for expanded or new mills between 1984 and 1994 [EPA. 1982]. The guidelines require costless chemical substitution for pentachlorophenol and trichlorophenol, and the costs for zinc removal are incorporated in the costs to comply with the Best Practicable Technology effluent guidelines.

6. The basis for the 1994 projection of production is explained in Appendix 3-B.

7. For 1973, we drew on three sources -- EPA's 1974 survey of mills [Wapora Inc. 1974], *The National Residuals Discharge Inventory* [Luken, Basta, and Pechan. 1976], and EPA's Development documents for the pulp and paper industry [EPA. 1974(a), 1974(b), and 1976] -- to estimate annual effluent loadings. For 1984, the procedure was more straight-forward. We just took the average of between 2 and 26 months of measured effluent data from each mill's Discharge Monitoring Reports. For 1994, we assumed that all mills out of compliance come into compliance, and that all existing mills that increase production and all new mills meet New Source Performance Standards.

8. In dealing with the compliance issue, we estimated costs based on the percent the mill is out of compliance. If the mill were only 10 percent out of compliance, then we assumed that its clarifier would be modified at 2 percent of the cost of the clarifier, and annual labor costs would increase 25 percent to bring that mill into compliance. If the mill were 20 percent out of compliance, we assumed the same modifications as if the mill were 10 percent out of compliance plus addition of a flocculent addition system. If the mill were 40 percent out of compliance, we assumed the same modifications as if the mill were 10 percent out of compliance and added a fine-screening system upstream of the clarifier. If the mill were 80 percent out of compliance, we replaced the entire clarification system with a new clarification system. Based on the percent of non-compliance for each non-complying mill, these assumptions were used to calculate compliance costs to the industry -- i.e., to bring each of the non-complying mills into compliance.

9. One potentially confounding factor in the water quality analysis is that 49 municipal wastewater treatment plants are located within 10 miles of 38 of the 68 mills. Their loadings with and without the Clean Water Act could alter the use characterization of the river segments. In order to check for this possibility, we simulated dissolved oxygen profiles for the 38 segments with only pulp and paper mill effluents. For 35 of the 38 segments, the miles of stream with dissolved oxygen less than 5.0 mg/l are identical or nearly identical to the miles in the simulation that combined pulp and paper mill and municipal wastewater treatment plant loadings. For the remaining three mills on three segments, the miles of stream with dissolved oxygen less than 5.0 mg/l are slightly different from the miles in the simulation that combined pulp and paper and municipal loadings. The

differences are not sufficient to change the use category that results from the combined loadings simulation, however.

10. The primary factors affecting the water quality results are the in-stream assimilative capacity of the receiving stream and its relationship to the loading reductions associated with the Clean Water Act. On the one hand, the 15 mills that discharge into rivers that remain in the unusable category (dissolved oxygen less than 3.5 mg/l) under both scenarios are located on rivers with dilution ratios less than 10. The low dilution ratios indicate a limited capacity to assimilate organic waste, particularly when the dilution ratio is less than one. Also, 5 of the 15 mills reduce their effluent by less than 50 percent as a result of the Clean Water Act, allowing limited potential for changes in in-stream dissolved oxygen. On the other hand, the 17 mills that discharge into rivers that remain in the superior game fishing category (dissolved oxygen greater than 5.0 mg/l) under both scenarios are located primarily on rivers with dilution ratios greater than 10. Higher dilution ratios indicate a greater capacity to assimilate organic waste, particularly when the dilution ratio is greater than 100. Lastly, the 36 mills that discharge into rivers that change potential use categories between the scenarios are located primarily on rivers with dilution ratios in the range of 1 to 100. Moderate dilution ratios offer the possibility that changes in effluent discharge can alter the dissolved oxygen profile of a river, but are not so low that even modest loadings would result in unsatisfactory water quality. (The three mills located on three segments with dilution ratios greater than 100 change only one use category, whereas most of the mills on the other segments change more than one use category.) Also, the 36 mills reduce their effluent by more than 50 percent as a result of the Clean Water Act. The larger reductions, particularly those in the 75-100 percent category, would normally result in changes in in-stream dissolved oxygen.

11. Our review provided limited information on pertinent qualitative characteristics of the 36 reaches whose activity classification changed as a result of the Clean Water Act. This information is summarized in Appendix 3-H. We found credible qualitative information only about access to the site and the nature of the recreation market. As can be seen, most potential recreation users are generally local residents. Only at six sites is the potential recreation population drawn from a larger region because of exceptional fishing or whitewater recreation opportunities. Many sites have limited access because of high banks and rugged terrain bordering the sites. Restricted local access, however, need not prohibit boating through the reach from upriver sites. Therefore, we do not rule out benefits because of local access problems. We could not

obtain quantitative information about market size in most cases and reliable information about the availability of substitute sites.

12. The choice of a 30-mile radius is consistent with several studies, including: (1) a University of Kentucky Water Resources Institute national survey of 3,000 fishermen, which reported that 92 percent of the sample traveled 30 miles or less to a site to fish [Bianchi. 1969], (2) an analysis of a 1982 Department of Interior survey, which indicated that the median distance traveled was 32 miles [geometric mean = 31.6 miles] to fish [Vaughan et al. 1985], (3) a National Boating Survey, which indicated that 66 percent of all boaters hauled their boats and that the mean distance to the site was 34 miles [U.S. Department of Transportation. 1979], and (4) a comprehensive survey of Wisconsin lake recreation, which found that the average distance traveled was 25 miles for combined day and overnight recreators [Parsons. 1989].

13. Naughton, Desvousges, and Parsons [1987] identify three major sources of bias in benefit transfers. Study error arises from inaccurate estimation in the original study from theoretical, econometric, or data deficiencies. Extrapolation error arises from extrapolation beyond the range of data used to estimate the study model. Correspondence error is attributable to uncontrolled differences in site characteristics between the study site and the new site.

14. Mitchell and Carson [1984] found that the annual household willingness to pay to improve the quality of the nation's lakes, rivers, and streams from nonboatable to swimmable was $275. We assigned an upper bound value of $90 for a similar improvement in just one stream. This supports the position that our upper bound value is a plausible value.

15. Assigning contingent valuation benefits to nonusers living within 30 miles of each site would approximately double the upper bound total benefits, but would result in net benefits for only one more site.

16. An unpublished empirical study by Harrington [1981] arrived at the same conclusion about the efficiency consequences of the Clean Water Act. Using data on recreation behavior in 25 cities and a crude measure of water quality improvements, he estimated the potential recreation benefits from compliance with the Best Practicable Technology policy. His estimate of annual recreation benefits is considerably less than the annual costs of compliance.

References

Bianchi, Dennis H. 1969. *The Economic Value of Streams for Fishing.*
Report No. 25. Water Research Institute, University of Kentucky,
Lexington, KY.

Brandes, Debra A. (ed.). 1984. *1985 Post's Pulp and Paper Directory.*
Miller Freeman Publications, San Francisco, CA.

Freeman, A. Myrick III. 1984. "On the Tactics of Benefit Estimation
Under Executive Order 12291" in *Environmental Policy Under Reagan's
Executive Order: The Role of Benefit-Cost Analysis*, V. Kerry Smith, ed.
UNC Press, Chapel Hill, NC.

Grayman, Walter M. 1985. "Routing and Graphical Display System
(RGDS) User's Manual." Report to the Monitoring and Data Support
Division, U.S. Environmental Protection Agency, Washington, DC.

_____. 1987. "Impacts of Pulp and Paper Industry on Water Quality."
Report to the Office of Policy Analysis, U.S. Environmental Protection
Agency, Research Triangle Institute, Research Triangle Park, NC.

Harrington, Winston. 1981. "The Distribution of Recreational Benefits
from Improved Water Quality: A Micro Simulation." Discussion Paper
D-80. Resources for the Future, Washington, DC.

Hazilla, Michael T. and Raymond J. Kopp. 1986. "The Social Cost of
Environmental Quality Regulations: A General Equilibrium Analysis."
Discussion Paper QE86-02. Resources for the Future, Inc., Washington,
DC.

Luken, Ralph A., Daniel J. Basta, and Edward H. Pechan. 1976. "The
National Residuals Discharge Inventory." National Research Council,
Washington, DC.

Medynski, Ann L. (ed.). 1973. *Post's 1974 Pulp and Paper Directory.*
Miller Freeman Publications, San Francisco, CA.

Mitchell, Robert Cameron, and Richard T. Carson. 1981. "An
Experiment in Determining Willingness to Pay for National Water
Quality Improvements." Draft report to the U.S. Environmental
Protection Agency, Resources for the Future, Inc., Washington, DC.

_____. 1984. *A Contingent Valuation Estimate of National Freshwater Benefits: Technical Report to the U.S. Environmental Protection Agency.* Resources for the Future, Inc., Washington, DC.

Naughton, Michael C., and William H. Desvousges. 1986. "Water Quality Benefits of Additional Pollution Control in the Pulp and Paper Industry." Report to the U.S. Environmental Protection Agency, Research Triangle Institute, Research Triangle Park, NC.

_____, William H. Desvousges, and George R. Parsons. 1987. "Benefits Transfer: Conceptual Problems in Estimating Water Quality Benefits Using Existing Studies." Unpublished Paper, San Diego State University, Economics Department.

Parsons, George R. 1989. Personal communication. Department of Economics, University of Delaware. Newark, DE.

Putnam, Hayes and Bartlett. 1986. "Cost of Clean Model of the Pulp, Paper and Paperboard Industry." Draft report to the Office of Policy, Planning, and Evaluation, U.S. Environmental Protection Agency. Cambridge, MA.

Smith, V. Kerry, and William H. Desvousges. 1986. *Measuring Water Quality Benefits.* Kluwer-Nijhoff Publishing, Boston, MA.

U.S. Bureau of the Census. 1986. "Pollution Abatement Costs and Expenditures." MA-200(84)-1. Washington, DC.

U.S. Environmental Protection Agency (EPA). 1974(a). "Development Document for Effluent Limitations Guidelines and New Source Performance Standards for the Unbleached Kraft and Semi-chemical Pulp Segment of the Pulp, Paper, and Paperboard Mills Point Source Category". EPA-440/1-74-025-a. Washington, DC.

_____. 1974(b). "Development Document for Effluent Limitations Guidelines and New Source Performance Standards for the Builders' Paper and Roofing Felt Segment of the Builders' Paper and Board Mills Point Source Category." EPA-440/1-74-026a. Washington, DC.

_____. 1976. "Development Document for Effluent Limitations Guidelines for the Bleached Kraft, Groundwood, Sulfite, Soda, Deink, and Non-Integrated Paper Mills Segment of the Pulp, Paper, and

Paperboard Point Source Category." EPA-440/1-76/047-b. Washington, DC.

_____. 1982. "Development Document for Effluent Limitations Guidelines and Standards for the Pulp, Paper, and Paperboard and the Builders' Paper and Board Mills Point Source Categories." EPA-440/1-82/025. Washington, DC.

_____. 1984. "The Cost of Clean Air and Water Report to Congress 1984." EPA-230/5-84-008. Washington, DC.

_____. 1986(a). "Ambient Water Quality Criteria for Dissolved Oxygen." EPA-440/5-86-003. Washington, DC.

_____. 1986(b). "Development Document for Best Conventional Pollutant Control Technology Effluent Limitations Guidelines and Standards for the Pulp, Paper, and Paperboard and the Builders' Paper and Board Mills Point Source Categories." EPA-440/1-86/025. Washington, DC.

U.S. Department of Interior. 1979. "Third Nationwide Outdoor Recreation Plan: The Assessment." Washington, DC.

U.S. Department of Transportation. 1979. "Recreational Boating in the Continental U.S. -- 1973 and 1976 Nationwide Boating Survey." U.S. Government Printing Office, Washington, DC.

U.S. President's Commission. 1987. *American Outdoors*. Island Press, Washington, DC.

Vaughan, William J., Charles M. Paulsen, Julia A. Hewett, and Clifford S. Russell. 1985. *The Estimation of Recreation-Related Water Pollution Control Benefits: Swimming, Boating, and Marine Recreational Fishing.* Report to the U.S. Environmental Protection Agency. Resources for the Future, Inc., Washington, DC.

Wapora, Inc. 1974. "Survey of Pulp and Paper Mills for 1974 and 1976 Development Documents." Memorandum to the Effluent Guidelines Division, U.S. Environmental Protection Agency. Rockville, MD.

Appendix 3-A

COSTS FOR INDIRECT-DISCHARGING MILLS

We started by compiling a 1984 inventory of 218 indirect-discharging mills from the records of EPA's Industrial Technology Division. Using the *Post's 1974 Pulp and Paper Directory* [Medynski. 1973], the *1985 Post's Pulp and Paper Directory* [Brandes. 1984], and the 1984 inventory, we estimated a population of 207 mills for 1973. Using our 1984-1994 projection of production (described in Appendix 3-B), we estimated that six new indirect-discharging mills would come on line between 1984 and 1994. The 1994 inventory of 224 mills is based on the assumption that none of the mills operating in 1984 ceased production.

We obtained the other data elements from the same sources we used to characterize the direct dischargers. The 1984 records for the Industrial Technology Division listed the subcategory of each mill. We based production levels for 1973 and 1984 on capacity figures from Post's directories, which had data on only 160 of the 207 mills in 1973 and 170 of the 218 mills in 1984. We took data on in-place treatment processes from Post's directories. If any treatment processes were listed, we assumed that internal controls were also in place. As a result, we could estimate pollution control costs for only 160 mills in 1973, 170 mills in 1984, and 176 mills in 1994.

We did not attempt to estimate effluent loadings from indirect discharging mills that are by definition connected to a municipal wastewater treatment plant. Data are not available to estimate their discharges.

The cost analysis for the indirect dischargers was much less complex than that for direct dischargers because we did not attempt to estimate compliance costs or the impacts of future regulations. The costs for indirect dischargers consist of treatment costs plus sewer charges paid to municipal governments. We estimated the treatment costs with the same cost functions used for the direct dischargers. We assigned sewer charges to mills based on the state sewer charges paid by the pulp and paper industry in 1973 and 1986 [U.S. Bureau of the Census. 1975, 1986]. First, we estimated the total production of the indirect dischargers in each state. Then we divided each state's production into the total sewer charges paid by mills in that state, resulting in annual sewer charges measured in dollars per ton of production.

Sewer charge data were not comprehensive. For some states sewer charges were not disclosed, usually because only one or two indirect mills were sampled in that state. The Census survey did not provide a number if the sewer charge total for the Standard Industrial Classification category in a state was under $50,000. For those states where this information was not disclosed or where the sewer charge was less than $50,000, a national average dollar per ton sewer charge was used. The 1973 national average was used for 16 states. The 1984 national average was used for 10 states.

For 1994, mill sewer charges were only modified based on the change in production between 1984 and 1994. Given the additional treatment requirements being imposed on municipal treatment systems, sewer charges will most likely increase between 1984 and 1994. Thus, 1994 sewer charges are probably underestimated.

Table 3-A-1 presents a national summary of water pollution control costs for indirect-discharging mills. The annual costs for these mills are between 10 and 15 percent of the combined total costs for direct- and indirect-discharging mills.

Table 3-A-1. National Summary of Water Pollution Control Costs for Indirect Discharging Mills in This Study ($1984 10^6)

Scenario	Capital	O&M	Sewer	Annual
1973[a]	$150	$15	$35	$70
1984 w/o CWA	$165	$15	$40	$75
1984 w/ CWA[a]	$360	$40	$80	$165
1984 with full compliance	N/A	N/A	N/A	N/A
1994	$385	$40	$85	$180

a. The pollution control cost per ton of production in 1973 was $8. It increased to $14 by 1984.
N/A = Not available.

References

Brandes, Debra A. (ed.). 1984. *1985 Post's Pulp and Paper Directory.*
Miller Freeman Publications, San Francisco, CA.

Medynski, Ann L. (ed.). 1973. *Post's 1974 Pulp and Paper Directory.*
Miller Freeman Publications, San Francisco, CA.

U.S. Bureau of the Census. 1975. "Pollution Abatement Costs and
Expenditures, 1973." MA-200(73)-1. U.S. Government Printing Office,
Washington, DC.

_____. 1986. "Pollution Abatement Costs and Expenditures, 1984."
MA-200(84)-1. U.S. Government Printing Office, Washington, DC.

Appendix 3-B

PULP AND PAPER MILLS EVALUATED IN THIS STUDY

1973 and 1984

The first step in the analysis was to establish an inventory of pulp and paper mills. We derived an initial 1984 inventory by checking EPA's Industrial Facilities Discharge file for all mills under the four relevant Standard Industrial Classification codes. We checked these mills against the *1985 Post's Pulp and Paper Directory* [Brandes. 1984] and eliminated all mills not operating in 1984. We then checked the inventory against the list of mills from EPA's Industrial Technology Division, which issued the pulp and paper effluent guidelines. We eliminated some mills because they were indirect dischargers and added others because, even though they were not found in the Industrial Facilities Discharge file, they were actually operating in 1984. At the end of our search, we had identified 334 directly discharging mills operating in 1984. Of these, we modeled 306 mills (Appendix B).

To produce a 1973 inventory, we started with our 1984 inventory. We checked that inventory against several sources -- EPA's 1974 survey of mills [Wapora. 1974], *The National Residuals Discharge Inventory* [Luken, Basta, and Pechan. 1976], and the *Post's 1974 Pulp and Paper Directory* [Medynski. 1973]. We subtracted mills that came on line between 1973 and 1984 and added mills that had closed between 1973 and 1984. The end result of our search identified 325 directly discharging mills operating in 1973. Of these, we modeled 306 mills (Appendix B).

The second step was to classify each mill by type on the basis of EPA's subcategorization scheme for the pulp and paper industry. EPA subdivided the industry into 25 subcategories that reflect how the industry operates with respect to raw materials, processing sequences, and product mix (Table 3-B-1). EPA used this subcategorization scheme to develop standardized effluent limitations for mills with similar characteristics.

The third step was to estimate production at each mill. For 1973 production, we used daily capacity estimates from the *Post's 1974 Pulp and Paper Directory*. For 1984, we took production data from the National

Table 3-B-1. EPA Subcategories: Pulp, Paper, and Paperboard Industry

Segment	Subcategory Code
Integrated Segment	
Dissolving Kraft	110
Market Bleached Kraft	120
BCT (Board, Coarse, and Tissue) Bleached Kraft	130
Alkaline Fine	140
Soda	140
Unbleached Kraft	
• Linerboard	150
• Bag and Other Products	450
Semi-Chemical	160
Unbleached Kraft and Semi-Chemical	170
Unbleached Kraft - Neutral Sulfite Semi-Chemical	160
Dissolving Sulfite Pulp	
• Nitration	211
• Viscose	212
• Cellophane	213
• Acetate	214
Papergrade Sulfite (Blow Pit Wash)	220
Papergrade Sulfite (Drum Wash)	230
Groundwood-Chemi-Mechanical	N/A
Groundwood-Thermo-Mechanical	320
Groundwood-CMN (Coarse, Molded, and News) Papers	330
Groundwood-Fine Papers	340
Secondary Fibers Segment	
De-ink	
• Fine Papers	1010
• Tissue Papers	1030
• Newsprint	1020
Paperboard from Wastepaper	1120
• Corrugating Medium Furnish	N/A
• Noncorrugating Medium Furnish	N/A
Tissue from Wastepaper	1110
Wastepaper-Molded Products	1130
Nonintegrated Segment	
Nonintegrated - Fine Papers	2010
• Wood Fiber Furnish	N/A
• Cotton Fiber Furnish	N/A
Nonintegrated Tissue Papers	2020
Nonintegrated Lightweight Papers	
• Lightweight Papers	2040
• Lightweight Electrical Papers	2060
Nonintegrated - Filter and Nonwoven Papers	2050
Nonintegrated - Paperboard	2110

N/A = Not applicable. EPA has no subcategory codes for these segments.

Pollutant Discharge Elimination System permits. We checked them with the *1985 Post's Pulp and Paper Directory* to eliminate obvious errors. In the case of mills in the South, we compared our production estimates with a Region IV study of the industry [EPA. 1986]. To compute annual production, we assumed that all mills operated 345 days per year. Actual days of operation in a given year are a function of economic conditions.

The fourth step was to estimate the pollution control equipment in place for the appropriate years. For 1973, we used three sources: EPA's 1974 survey of mills [Wapora. 1974], *The National Residuals Discharge Inventory* [Luken, Basta, and Pechan. 1976], and the *Post's 1974 Pulp and Paper Directory* [Medynski. 1973]. If no pollution control equipment was listed in any of the sources, we assumed discharge to surface water. If any equipment was in place, we assumed internal controls were in place because standard industry practice is to install internal controls before installing external treatment. For 1984, we used a more straightforward procedure. We just took the information from the discharge permit for the mill. However, we did check with the *1985 Post's Pulp and Paper Directory* to eliminate any obvious errors or omissions.

The fifth step was to establish the unit effluent limitation (pounds of pollutant per ton of production) for each mill in 1984. For most mills, we obtained that data from their discharge permit. For those few mills with no permit limitation, we assumed that the permit limit equaled the subcategory limit in the effluent guidelines [EPA. 1982].

1994

The American Paper Institute [API. 1987] has announced expansions and additions for the period 1985-1990. For a few mills production is expected to decrease; but for many mills production is projected to increase by 1994. API's projection falls into two categories: those where the subcategory or subcategories of the mill will change from 1984 to 1994, and those where they will not change. API also projected nine new direct-discharging mills to come on line by 1994.

In addition to API's projections, we estimated further capacity and production increases based on projected GNP growth of 2.5 percent to 1994 and the relation between GNP and pulp and paper production. We distributed this additional capacity first among mills that API projected to grow and second among existing mills in regions where expansions are likely. To achieve the estimated total capacity, we added five more new directly discharging mills to come on line in addition to the nine API projected.

Treatment processes in place for mills in 1994 with no projected production changes were assumed to be the same as in 1984, if the subcategory of the mill does not change. For existing mills with either production or subcategory changes and for new mills, we assigned treatment processes that would meet New Source Performance Standards.

References

American Paper Institute (API). 1987. *Paper Paperboard Wood Pulp Capacity*. New York, NY.

Brandes, Debra A. (ed.). 1984. *1985 Post's Pulp and Paper Directory*. Miller Freeman Publications, San Francisco, CA.

Luken Ralph A., Daniel J. Basta, and Edward H. Pechan. 1976. *The National Residuals Discharge Inventory*. National Research Council, Washington, DC.

Medynski, Ann L. (ed.). 1973. *Post's 1974 Pulp and Paper Directory*. Miller Freeman Publications, San Francisco, CA.

Wapora, Inc. 1974. "Survey of Pulp and Paper Mills for 1974 and 1976 Development Documents." Memorandum to the Effluent Guidelines Division, U.S. Environmental Protection Agency. Rockville, MD.

U.S. Environmental Protection Agency (EPA). 1982. "Development Document for Effluent Limitation Guidelines and Standards for the Pulp, Paper and Paperboard and Builders' Paper and Board Mill Point Source Categories." EPA-440/1-82/025. Washington, DC.

_____. 1986. "Study of the Pulp and Paper Industry in Region IV." Atlanta, GA.

Appendix 3-C

WATER POLLUTION CONTROL COST ESTIMATION

We used 14 specific cost functions for each of the 25 subcategories (plus subdivisions of subcategories) identified by EPA in Table 3-B-1 [E.C. Jordan. 1985]. For each subcategory, we selected treatment components from Table 3-C-1 to formulate a treatment system that would enable a mill to comply with Best Practicable Technology or Best Conventional Technology effluent levels. With a few exceptions, the alphanumeric codes of the treatment functions match the codes in the discharge permit applications.

Three types of costs -- capital, operation and maintenance, and annual -- were computed in each year for each in-place treatment process. The first two types of costs, capital and operation and maintenance, were computed using engineering cost functions for the treatment processes which are subcategory-specific and of the standard form:

$$\text{Cost} = a * X^b$$

where: X = the average daily production of the mill in tons of paper (and paperboard and market pulp);

a = a linear scaling factor;

b = a scaling factor for economies of scale in pollution control technology.

The annual costs were computed by adding the annual operation and maintenance costs to the annualized capital costs, based on amortizing the capital costs over 15 years at a 10 percent discount rate, yielding a capital recovery factor of 0.1315.

Reference

Jordan, E.C. (Consulting Engineers). 1985. "MMBA: Water Cost Functions for the Pulp & Paper Industry." Draft report to the Office of Policy, Planning and Evaluation, U.S. Environmental Protection Agency. Portland, ME.

Table 3-C-1. Treatment Components

PHYSICAL TREATMENT PROCESSES

1-I. Foam Fractionation	1-T. Screening	1-U. Sedimentation (Settling)

BIOLOGICAL TREATMENT PROCESSES

3-A. Activated Sludge	3-B. Aerated Lagoons

OTHER PROCESSES

4-A. Discharge to Surface
Water

SLUDGE TREATMENT AND DISPOSAL PROCESSES

5-J. Flotation Thickening	5-T. Sludge Lagoons	5-Y. Landfill of Combined
5-O. Incineration[a]	5-U. Vacuum Filtration[a]	Primary and
5-Q. Landfill[a]	5-X. Vacuum Filtration of	Secondary Sludge
	Combined Primary	5-Z. Incineration of
	and Secondary	Combined Primary
	Sludge	and Secondary Sludge
	(i.e., if 1-U and 3-A	
	or 3-B in place)	

Source: NPDES Permit Application, Table 2 C-1 (we added 5-X, 5-Y, and 5-Z).

a. Primary sludge only.

Note: The processes listed above are only a small fraction of the universe of treatment processes available. They represent those processes for which accurate cost information was available.

Appendix 3-D

WATER POLLUTION CONTROL COST COMPARISONS

We compared our real resource cost estimates from the Cost of Clean model with other real resource and financial cost estimates. Other real resource cost estimates are in the EPA "Cost of Clean Air and Water Report to Congress 1984" [1984] and in the EPA "Development Document for Best Conventional Technology" [1986] for the pulp and paper industry. The two estimates of financial costs are the Census' "Pollution Abatement Costs and Expenditures, 1984" [U.S. Bureau of the Census (Census). 1986] and the paper industry's "A Survey of Pulp and Paper Industry Environmental Protection Expenditures -- 1984" [National Council of the Paper Industry for Air and Stream Improvement, Inc. 1985]. We did not include the paper industry survey in our comparison because it did not report operation and maintenance or annualized capital costs; it reported only capital investments.

The "Cost of Clean Air and Water Report to Congress 1984" estimated the annual capital, operation and maintenance, and annual costs for the pulp and paper industry. These costs were obtained by updating for inflation a 1977 EPA consultant report, "Economic Impacts of Pulp and Paper Industry Compliance with Environmental Regulations" [Arthur D. Little. 1977]. These cost estimates were updated to 1984 dollars and are presented in Table 3-D-1 as EPA "Cost of Clean." The annual costs for water pollution control are relatively comparable to the Cost of Clean model costs; the major difference is the annual capital cost.

The 1986 Development Document for the Best Conventional Technology effluent guidelines includes an estimate of real resource costs for compliance with Best Practicable Technology effluent guidelines. The annual capital and operation and maintenance costs were adjusted to 1984 dollars and the annual capital costs were annualized with a capital recovery factor reflecting a 10 percent cost of capital and a 15-year life for pollution control equipment rather than a 17 percent cost of capital and a 10-year life for pollution control equipment.

We matched 220 mills from the Cost of Clean model and the 1986 Development Document in order to have a comparable cost estimate. The

Table 3-D-1. Aggregate Cost Comparisons ($1984 10⁶)

Cost Category	EPA	Model[a]	Census[b]
Capital In-Place	-	$4,490	-
Investment	$10	-	$60
Annual Capital	$490	$590	$120
Annual O&M	$450	$430	$360
Annual Costs	$940	$1,020	$480

a. Direct-discharging mills only. They account for approximately 85 percent of the total annual costs derived from the Cost of Clean model.
b. Census investment and annual cost figures from Table 4A, annual capital and annual O&M figures were interpolated from Table 5A.

annual costs of compliance with the Best Practicable Technology effluent guidelines in the 1986 Development Document are $930 million. The annual costs of compliance with Best Practicable Technology in the Cost of Clean model are $680 million for the same 220 mills. The 1986 Development Document annual cost estimates are higher in the aggregate than the Cost of Clean estimates. They are higher for the integrated and nonintegrated subcategories and lower, except in one case, for the secondary fiber subcategories.

As a result of this comparison of real resource cost estimates, the Cost of Clean model estimate of pollution control costs appears to be a reasonable cost to compare to the benefits of water quality improvements. The Cost of Clean model estimate is equivalent to the 1984 Cost of Clean Air and Water estimate and approximately 75 percent of the 1986 Development Document estimate.

The Cost of Clean model real resource cost estimates, however, are higher than the financial costs of pollution control to the industry according to the Census data. Financial costs reflect how tax provisions, such as accelerated depreciation, reduce the actual costs of pollution control to industry. The difference between the annual capital cost estimate in the Cost of Clean model and the annual capital cost estimate as reported by the Census accounts for most of the difference between the annual real resource costs and financial costs. The annual operation and maintenance costs from the Cost of Clean model and the Census survey are not significantly different because there is little potential for tax distortions.

References

Arthur D. Little. 1977. "Economic Impacts of Pulp and Paper Industry Compliance with Environmental Regulations." EPA-230/3-76-014-1. Cambridge, MA.

National Council of the Paper Industry for Air and Stream Improvement, Inc. (NACASI). 1985. "A Survey of Pulp and Paper Industry Environmental Expenditures -- 1984." Special Report No. 85-04. New York, NY.

U.S. Bureau of the Census. 1986. "Pollution Abatement Costs and Expenditures." MA-200 (84)-1. U.S. Government Printing Office, Washington, DC.

U.S. Environmental Protection Agency. 1984. "Cost of Clean Air and Water Report to Congress 1984." EPA-230/5-84-008. Washington, DC.

_____. 1986. "Development Document for Best Conventional Pollutant Control Technology Effluent Limitations Guidelines for the Pulp, Paper and Paperboard and the Builders' Paper and Board Mills." EPA-440/1-86-025. Washington, DC.

Appendix 3-E

LOADINGS DATA

We characterized the effluent loadings from pulp and paper mills using two different information sources. We took the BOD data for all three scenarios for the 68 pulp and paper mills from the Cost of Clean model. They are presented in Table 3-E-1. The wastewater flow, available only for 1984, came from EPA's Industrial Facilities Discharge file [EPA. 1987]. The amount of wastewater flow for the other two scenarios -- 1973 and 1984 without the Clean Water Act -- was assumed equal to the 1984 flow. The flow data are used in determining the immediate dissolved oxygen balance in the stream at the discharge site. All scenarios assume that there was no dissolved oxygen in the effluent.

We characterized the effluent loadings from municipal sources in 1973 and 1984 by using two different approaches. In estimating municipal loadings for 1973, EPA's "Needs Survey" [EPA. 1973] and engineering assumptions provided BOD loadings and population served, and a combination of the Industrial Facilities Discharge file and engineering assumptions provided the flow data. In estimating municipal loadings for 1984, EPA's "Needs Survey" [EPA. 1984] provided BOD loadings and population served, and the Industrial Facilities Discharge file provided flow values. In estimating municipal loadings for 1984 without the Clean Water Act, the BOD loadings per capita came from the 1973 estimates, and the population served came from the 1984 "Survey of Needs." The results of these calculations are presented in Table 3-E-2.

References

U.S. Environmental Protection Agency (EPA). 1973. "Report to Congress: Costs of Construction of Publicly Owned Wastewater Treatment Works -- 1973 Needs Survey." Washington, DC.

_____. 1984. "1984 Needs Survey -- Report to Congress." EPA-430/9-84-001. Washington, DC.

_____. 1987. "Industrial Facilities Discharge File." Computer Listing. Washington, DC.

Table 3-E-1.   Effluent Loadings Data for Pulp and Paper Mills
Analyzed in Water Quality Simulations

Mill	BOD$_5$ Loadings (pounds/day) 1973	1984 w/o CWA	1984 w/ CWA	Mill	BOD$_5$ Loadings (pounds/day) 1973	1984 w/o CWA	1984 w/ CWA
14	3,900	4,720	3,790	159	6,000	6,780	6,750
16	26,320	49,630	6,200	165	28,110	48,840	13,420
21	15,400	17,950	11,150	167	22,750	21,700	3,370
22	20,000	38,880	5,140	169	1,120	1,050	50
39	5,100	6,910	3,890	170	770	870	60
42	3,640	3,930	2,550	182	240	150	80
43	7,340	10,760	7,780	187	580	580	130
44	8,000	9,870	6,740	192	970	1,040	670
48	63,550	84,130	3,480	193	100	50	30
51	13,780	14,330	540	194	2,250	2,020	150
56	24,840	81,950	58,800	198	10,400	9,130	2,340
57	10,000	11,900	11,150	201	5,100	5,020	4,160
66	960	960	130	202	140	170	130
77	4,270	7,940	6,610	206	140	250	50
79	8,520	8,300	7,270	207	1,920	1,738	560
82	21,270	45,730	6,680	209	40	40	40
85	7,700	8,710	2,190	216	11,360	12,380	5,020
87	1,400	12,160	5,210	217	9,920	16,680	12,890
89	24,610	41,100	12,670	219	12,402	13,530	2,570
91	34,300	45,010	8,770	235	9,270	7,370	3,200
92	36,000	32,400	660	238	9,070	16,040	7,200
97	1,200	1,200	260	260	740	1,040	960
100	29,160	26,570	2,190	280	5,000	5,500	2,550
102	104,000	188,800	12,430	286	1,380	1,380	440
103	21,000	21,870	8,080	287	30	80	50
105	19,340	40,330	7,360	288	3,200	4,000	730
106	520	360	140	289	5,860	6,410	1,410
125	100	140	140	293	9,340	9,830	6,870
130	2,290	2,050	60	295	3,020	3320	2,200
141	2,310	1,390	120	296	29,700	32,500	5,540
147	15,130	38,500	3,900	324	7,000	7,000	260
150	10,350	17,250	2,100	343	29,480	51,188	2,330
151	2,650	3,570	450	344	1,800	1,950	170
157	540	750	30	355	243,000	291,600	2,400

Table 3-E-2.   Effluent Loadings Data for Municipal Treatment Plants
Analyzed in Water Quality Simulations

Plant	BOD$_5$ Loadings (lbs/day) 1973[a]	1984 w/o CWA	1984 w/ CWA	Plant	BOD$_5$ Loadings (lbs/day) 1973[a]	1984 w/o CWA	1984 w/ CWA
1001	240	310	30	1026	20	20	20
1002	0	0	80	1027	40	40	40
1003	1,220	2,940	900	1028	20	20	10
1004	80	240	10	1029	10	40	30
1005	110	270	270	1030	210	270	210
1006	1,250	1,350	550	1031	700	700	80
1007	0	0	90	1032	30	50	30
1008	970	970	970	1033	2,820	3,080	290
1009	1,320	1,470	1,470	1034	290	280	50
1010	80	220	220	1035	20	60	30
1011	10	10	10	1036	30	60	60
1012	80	130	130	1037	0	0	0
1013	1,810	1,500	320	1038	90	90	90
1014	50	80	30	1039	0	0	10
1015	80	80	10	1040	10	50	0
1016	1,230	1,660	90	1041	70	80	20
1017	0	0	30	1042	60	100	90
1018	90	320	90	1043	30	50	50
1019	480	960	170	1044	150	130	130
1020	0	0	140	1045	150	1,250	40
1021	900	2,110	200	1046	20	310	310
1022	10	30	20	1047	10	10	10
1023	13,160	5,920	1,520	1048	20	240	240
1024	20	20	20	1049	20	70	70
1025	10	10	10				

a. In estimating municipal loadings for 1973, the following hierarchy of rules was applied:

Available Data	Estimation Method
BOD loading and flow	Use available values
Population and treatment type	Assume 100 gal/capita/day and use matrix of BOD concentration by treatment type
Treatment Type	BOD Concentration (mg/l)
No Treatment	200
Primary Treatment	140
Lagoon	25
Oxidation Ponds	25
Trickling Filter	25
Population only	Assume primary treatment and 100 gal/cap/day
No data	Assume same flow and loading as 1984
"BOD not controlled"	Assume BOD=200 mg/l; 1984 flow value
Not operating in 1973	BOD=O; Flow=O
BOD in lbs/day; no flow data	Use given BOD; assume 1984 flow

Appendix 3-F

WATER QUALITY MODEL

This appendix reviews the classical Streeter-Phelps BOD - dissolved oxygen (D.O.) formulation along with the details of its application in this study. It also summarizes the results of the sensitivity analysis of the resulting model predictions to variations in model parameters.

In the Streeter-Phelps formulation, BOD is represented as a first-order decay function of the form:

$$L = L_o e^{-K_1 t} \qquad \text{(Eq.1)}$$

where:
- L = ultimate BOD at time t;
- L_o = BOD at time 0;
- K_1 = deoxygenation coefficient (per day) ;
- t = time in days

The dissolved oxygen deficit, D, is the difference between the saturation dissolved oxygen concentration and the dissolved oxygen concentration, and is calculated by the following equation:

$$D = [K_1 L_o/(K_2 - K_1)] (e^{-K_1 t} - e^{-K_2 t}) + (D_o e^{-K_2 t}) \quad \text{(Eq. 2)}$$

where:
- D = dissolved oxygen deficit (mg/l);
- K_2 = reaeration coefficient (per day);
- D_o = initial dissolved oxygen deficit;

The saturation concentration of dissolved oxygen in mg/l may be calculated by the following formula:

$$\text{DOSAT} = 14.62 - 0.367\, T + 0.0045\, T^2 \qquad \text{(Eq. 3)}$$

where:
- T = temperature (°C) and the salinity concentration is assumed to be zero.

Experimental and theoretical work has shown that the reaeration coefficient may generally be estimated based on physical stream characteristics. Two commonly used formulations for predicting reaeration are:

$$K_2 = a\ U^b / H^c \qquad\qquad \text{(Eq. 4)}$$

and $\quad K_2 = d\ (US)^f / H^g \qquad\qquad \text{(Eq. 5)}$

where: $\quad$ H $\;=\;$ mean depth (feet);
$\qquad\qquad$ U $\;=\;$ mean velocity (feet per second);
$\qquad\qquad$ S $\;=\;$ stream slope (feet per feet);
$\quad$ a,b,c,d,f,g $\;=\;$ fitting coefficients.

In this study the former formulation, using velocity and depth, is used because of the availability or derivability of these parameters from the information in the Reach File. Seven different equations based on this formulation were studied. The potential coefficients (a, b, and c) for this formulation are presented in Table 3-F-1.

Table 3-F-1. Reaeration Coefficients

Experimenters	a	b	c
O'Connor-Dobbins [1958]	12.9	0.5	1.5
Churchill, Elmore, Buckingham [1962]	11.56	0.969	1.673
Owens, Edwards, Gibbs [1964]	25.07	0.73	1.75
Langbein-Durum [1967]	7.59	1.00	1.33
Negulescu-Rojanski [1969]	10.9	0.85	0.85
Issacs-Gaudy [1966]	8.6	1.00	1.5
Bennett-Rathbun [1972]	20.15	0.607	1.689

The deoxygenation coefficient (K_1) is generally assumed to relate more to the nature of the BOD in the stream rather than the physical characteristics of the stream. A typical value for the deoxygenation coefficient for domestic waste is 0.23 per day [Fair and Geyer. 1958]. Sensitivity analysis was applied to analyze the effects of variation in K_1.

These equations have all been adjusted for an assumed temperature of 20°C and for use with the natural logarithmic form of the Streeter-

Phelps equations. The reaeration and deoxygenation coefficients may be adjusted for use at temperatures other than 20°C through the following relationship:

$$(K_2)_T = (K_2)_{20} \times 1.047^{T-20} \qquad \text{(Eq. 6)}$$

When applying the model in the Reach File - RGDS context, several application-specific assumptions are needed. These assumptions and methods are detailed below.

Within the Reach File, stream flows are stored at the downstream end of each reach corresponding to estimated average flow, low flow (approximation of the 7-day/10-year low flow), and average monthly flows for the 12 months. For this study, low flow and average summer stream flow, corresponding to an average of the streamflows for July, August, and September, as stored in the Reach File were used in the analysis. Velocity was based on the RGDS velocity relationship which uses stream flow and hydrologic stream order to determine a representative stream velocity [W.E. Gates. 1982].

In order to estimate stream depths, a general relationship was used to estimate the stream width at average flow [Leopold and Maddock. 1953]:

$$\text{Width}_{av} = 4 \times Q_{av}{}^{0.5} \qquad \text{(Eq. 7)}$$

where width is in feet and Q_{av} , the average stream flow, is in cubic feet per second (cfs). The width is then adjusted to reflect the summer flow, Q by the following relation:

$$\text{Width}_{summer} = (Q/Q_{av})^{0.5} \qquad \text{(Eq. 8)}$$

By continuity, stream depth is then calculated as:

$$\text{Depth} = (Q/V) / \text{Width}_{summer} \qquad \text{(Eq. 9)}$$

where Depth is in feet, Q is in cfs, and Velocity is in feet per second. Similar adjustments were made to reflect low flow conditions.

Within the Reach File, stream flow estimates are available at the downstream end of each reach. In this study, it is assumed that this flow may be applied for the entire reach. This is shown schematically in Figure 3-F-1. Furthermore, it is assumed that the flow estimates, which are based

Figure 3-F-1. Streamflow Representation Used in This Study

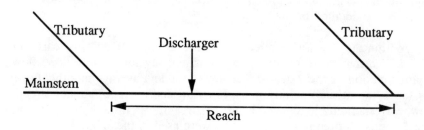

on streamgauge records, include the flow impacts of the discharges, and thus flows are not increased at the discharger locations. In calculating the dilution effects of the receiving stream on the discharges, the following formulations are used:

$$BOD_{DS} = BOD_{US} + (BOD_{dis}/Q_r)/5.39 \qquad \text{(Eq. 10)}$$

and $\quad D.O_{DS} = [(D.O._{US} \times Q_r) + (D.O._{dis} \times Q_{dis})]/(Q_r + Q_{dis})$
$$\text{(Eq. 11)}$$

where:

$D.O._{US}$ and BOD_{US} are the dissolved oxygen and BOD upstream of the discharger in the stream in mg/l;

$D.O._{DS}$ and BOD_{DS} are the dissolved oxygen and BOD immediately downstream of the discharger in the stream in mg/l assuming complete and instantaneous mixing;

Q_r is the stream flow upstream of the discharger in cfs;

Q_{dis} is the plant discharge in cfs;

5.39 is a factor for converting concentration from pounds per day per cfs to mg/l;

D.O.$_{dis}$ is the dissolved oxygen concentration in the discharge in mg/l; and

BOD$_{dis}$ is the ultimate BOD discharged from the mill in pounds/day.

Thus, in this formulation, BOD and dissolved oxygen from the mill discharge are being added without changing the value of the flow in the stream downstream of the plant. However, the relative magnitude of flows in the stream and from the discharger are considered in performing a mass balance for dissolved oxygen. Because BOD mill loadings are generally reported in terms of 5-day BOD, the following equation is used to convert 5-day BOD to ultimate BOD.

$$BOD_{ult} = 1.46 \times BOD_5 \qquad \text{(Eq. 12)}$$

where both ultimate and 5-day BOD are in mg/l.

Since there is both uncertainty in defining localized parameters used in the modeling and stochastic variation in many of the variables, a sensitivity analysis was performed to determine the variation in results under circumstances differing from the "base conditions." The most significant variable was found to be stream flow. As flow was decreased from the average summer conditions down to a low that would be expected to occur once in 10 years, the amount of stream miles impacted increased very significantly as did the difference between 1973 and 1984 conditions.

The sensitivity analysis indicated some variations in impacts due to changes in other parameters. Both the reaeration and deoxygenation coefficients affected results; although these variations were significantly less important than those associated with stream flow. Similarly, stream temperature as it affects these coefficients and the saturation dissolved oxygen level impacted the predicted results. Background conditions (dissolved oxygen and BOD) affected the absolute miles of streams impacted but had little effect on the relative impacts between 1973 and 1984.

References

Bennett, James P., and Ronald E. Rathbun. 1972. "Reaeration in Open-Channel Flow." Professional Paper 737. U.S. Geological Survey, Washington, DC.

Churchill, M.A., H.L. Elmore, and R.A. Buckingham. 1962. "The Prediction of Stream Reaeration Rates." *American Society of Chemical Engineers*, SA4, 88:1-46.

Fair, Gordon M., and John C. Geyer. 1958. *Elements of Water Supply and Waste-Water Disposal.* John Wiley & Sons, Inc., New York, NY.

Issacs, W.P., and A.F. Gaudy. "Atmospheric Oxygenation in a Simulated Stream." *American Society of Chemical Engineers*, SA2, 94:319-344.

Langbein, Walter B., and Walthon H. Durum. "The Aeration Capacity of Streams." Circular 542. U.S. Geological Survey, Washington, DC.

Leopold, Luna B., and Thomas Maddock, Jr. 1953. "The Hydraulic Geometry of Stream Channels and Some Physiographic Implications." Professional Paper 252. U.S. Geological Survey, Washington, DC.

Negulescu, M., and V. Rojanski, 1969. "Recent Research to Determine Reaeration Coefficient." *Water Research*, 3:189-202.

O'Connor, Donald J., and W.E. Dobbins. 1958. "Mechanisms of Reaeration in Natural Streams." *Transactions of the American Society of Chemical Engineers*, 123:641-684.

Owens, M., R.W. Edwards, and J.W. Gibbs. 1964. "Some Reaeration Studies in Streams." *International Journal of Air and Water Pollution*, 8:469-486.

W.E. Gates and Associates, Inc. 1982. "Reach File Enhancements: Pollutant Routing and Graphical Display Capabilities." Report to the Office of Water Regulations and Standards, U.S. Environmental Protection Agency. Fairfax, VA.

Appendix 3-G

MODEL VALIDATION

This appendix summarizes an assessment of the validity of the water quality modeling in this study [Grayman. 1987]. The assessment compared the results of the RGDS modeling effort to data available in EPA's STORET system. Within STORET, much of the data are keyed to the reach numbering system in use by EPA for cataloging stream-related information. Thus, it was possible to retrieve water quality data selectively in terms of the location of pulp and paper mills which are referenced to the reach numbering system.

Specifically, water quality data were retrieved that were located on a direct flow path within 20 miles downstream of each of the 68 pulp and paper mills. These retrievals were limited to dissolved oxygen measurements during the summer period (July through September). This period was selected because of the generally lower flows and higher temperatures which occur at that time. These conditions are most closely associated with the low flow conditions used in the modeling effort.

The water quality data were retrieved for two periods: 1970-1975 and 1980-1985. The former period was considered to be representative of the 1973 modeling scenario, whereas the latter period reflects the 1984 scenario. Six-year periods were used in each case to allow the retrieval of a sufficiently large amount of data to perform the required statistical analyses. Data were available for only 27 of the 68 mills for the 1970-1975 and the 1980-1985 periods.

Summary of the Results

To assess the results of this analysis, the change in *predicted* dissolved oxygen from 1973 to 1984 and the change in average *observed* dissolved oxygen from 1970-1975 to 1980-1985 were compared for each mill. Figure 3-G-1 summarizes the comparison. In this figure, the change in dissolved oxygen based on observed data is plotted against the modeled change in dissolved oxygen. Each point corresponds to a single mill.

Figure 3-G-1. Dissolved Oxygen Change 1973 - 1984

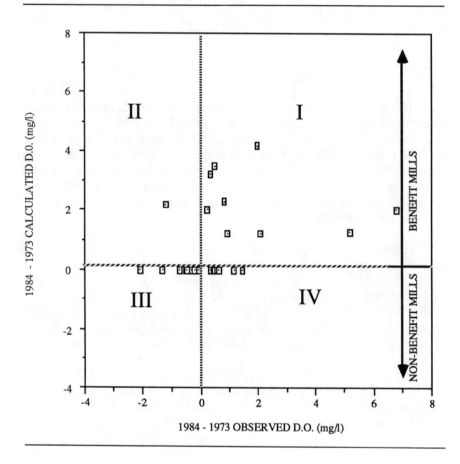

Referring to Figure 3-G-1, the results of the assessment suggest the following conclusions:

1. Benefit mills are those for which calculated dissolved oxygen increased from 1973 to 1984. These mills are shown in quadrants I and II of the figure. Of the 10 benefit mills for which data were available, nine showed increases in dissolved oxygen from the 1970s to the 1980s (points in quadrant I) and only one showed a decrease in dissolved oxygen (quadrant II). For the nine mills that showed an increase, the average observed increase was 2.07 mg/l, whereas the calculated average increase for the same mills was 2.10 mg/l. This strong correlation between observed

and calculated results suggests that those mills identified as benefit mills by the modeling process did indeed exhibit improvements in water quality.

2. Non-benefit mills are those mills which experience little or no improvements in water quality from 1973 to 1984 based on the modeling. These mills are shown in quadrants III and IV (these mills actually lie on the x-axis because no decreases in dissolved oxygen level were predicted by the modeling from 1973 to 1984). For all 17 non-benefit mills for which field data were available, the average net change in dissolved oxygen from the 1970s was either negative or a positive value of less than 1.45 mg/l. Seven of these mills experienced an increase in dissolved oxygen averaging 0.73 mg/l, whereas 10 experienced a decrease averaging 0.83 mg/l. The relatively modest change in dissolved oxygen evenly distributed between increases and decreases for these non-benefit mills strongly suggests that the non-benefit mills, as predicted by the modeling, did in fact experience little improvement in water quality as measured by dissolved oxygen.

Limitations of the Analysis

There are some inherent limitations to this analysis. The primary limitation is the availability of data. Dissolved oxygen data were available (within 20 miles downstream) for only about half of the mills in either of the two time periods and for only about 40 percent of the mills in both time periods. The amount of data also varied significantly from only a few data points for some mills to several hundred for other cases. Furthermore, unlike a modeling situation where the modeler can control the parameters (e.g., select low flow cases), we did not know whether the samples were spatially representative of the receiving stream below the mills nor whether they corresponded to relatively low flow periods. Thus, the reader should consider the results of the analysis as generally supportive rather than statistically defensible.

Reference

Grayman, Walter, M. 1987. "Investigation of the Validity of the Modeling Results Applied to Pulp and Paper Mills Based on Field Data." Report to the Office of Policy, Planning, and Evaluation, U.S. Environmental Protection Agency. Research Triangle Institute, Research Triangle Park, NC.

Appendix 3-H

VISITATION ESTIMATES

Table 3-H-1. Data for Calculating Recreation Participation Rates

Mill	Site Access	User Population	30-Mile Hslds.[a]	Stream Availability Ratio[b]	Assumed Participation Rate[c]
14		local	100,000	45	0.1
16		local	26,000	6	0.5
21		local	35,000	3	0.5
22	good	regional	172,000	4	d
39		local	125,000	23	0.1
42		local	76,000	8	0.5
43		local	14,000	5	0.5
44		local	199,000	44	0.1
48	limited	local	76,800	3	0.5
51	limited	local	19,300	16	0.1
56		local	130,000	4	0.5
57		local	34,000	5	0.5
66	limited	local	275,900	5	0.5
77		local	33,000	10	0.3
79		local	39,000	4	0.5
82		local	83,000	6	0.5
85		local	90,000	6	0.5
87		local	69,000	3	0.5
89		local	170,000	117	0.1
91	limited	local	7,100	5	0.5
92		local	154,000	10	0.5
97		local	123,000	162	0.1
100	good	regional	32,000	6	d
102	adequate	regional	32,000	67	d
103	adequate	local	52,000	7	0.5
105	adequate	local	7,000	10	0.3
106		local	152,000	8	0.5
125		local	292,000	102	0.1
130	limited	local	324,900	12	0.3
141		local	21,000	63	0.1
147	limited	regional	2,600	7	d
150		local	6,000	153	0.1
151		local	32,000	8	0.5
157		local	42,000	14	0.3
159	adequate	local	56,000	8	0.3

(cont.)

Table 3-H-1. (cont.) Data for Calculating Recreation Participation Rates

Mill	Site Access	User Population	30-Mile Hslds.[a]	Stream Availability Ratio[b]	Assumed Participation Rate[c]
165	adequate	local	31,000	33	0.5
167	limited	local	29,500	19	0.1
169	adequate	local	183,000	7	0.5
170		local	172,000	4	0.5
182	limited	local	324,900	13	0.3
187		local	51,000	63	0.1
192		local	149,000	27	0.1
193		local	70,000	16	0.1
194	good	local	324,900	14	0.3
198	limited	local	13,800	3	0.5
201		local	192,000	11	0.3
202		local	322,000	16	0.1
206		local	285,000	e	0.3
207		local	292,000	51	0.1
209		local	272,000	e	0.3
216		local	129,000	6	0.5
217		local	49,000	25	0.1
219	limited	local	132,500	40	0.5
235	limited	local	69,300	8	0.5
238	limited	local	23,300	6	0.5
260		local	256,000	9	0.5
280		local	48,000	e	0.3
286		local	50,000	6	0.5
287		local	54,000	115	0.1
288	limited	local	14,800	13	0.3
289	good	local	29,500	4	0.5
293		local	56,000	8	0.5
295		local	95,000	16	0.1
296		local	140,000	12	0.3
324		local	59,000	9	0.5
343	good	regional	17,000	5	0.5
344	good	regional	19,200	5	0.5
355	good	local	56,800	1	0.5

a. Number of households in 1984 within a 30-mile radius of a mill.
b. Total stream miles within 30-mile radius divided by stream miles affected by mill discharges.
c. Assumed participation rates based on stream availability ratios as follows:

Ratio	Assumed Substitutes	Participation Rate
1-9	Few	0.5
10-15	Some	0.3
>15	Many	0.1

d. Visitor use information available. Assumed participation rate not used.
e. Stream availability information not available. Mid-range participation rate assumed.

Appendix 3-I

RECREATION BENEFIT STUDIES

We cannot generally observe a market price for water-based recreation. Therefore, it is necessary to obtain estimates of willingness to pay either by means of a questionnaire that asks respondents directly (the contingent valuation method) or by means of observed travel behavior and associated costs (the travel cost method). People may enjoy benefits from improved water quality because they actually use the site, because they may use it in the future, or because they enjoy knowing that water quality has improved even though they do not expect to visit the site themselves. The travel cost method estimates only the willingness to pay of current users. The contingent valuation method can incorporate the willingness to pay of both users and nonusers. (For a discussion of contingent valuation method applications to water resources, see Brookshire, Eubanks, and Sorg [1986].)

There are eight available water quality benefit studies that are potential candidates for transferring values. Five of these are contingent valuation method studies: Walsh et al. [1978], Sutherland and Walsh [1985], Mitchell and Carson [1984], Gramlich [1977], and Smith and Desvousges [1986]. Two are travel cost method studies: Vaughan and Russell [1982(a)] and Smith, Desvousges, and Fisher [1984]. The eighth study is only a user participation study by Vaughan and Russell [1982(b)]. Naughton et al. [1986, 1987] argue that five of these studies can be eliminated because they are too dissimilar to the transfer sites, which are generally eastern rivers with local recreation use. Walsh et al. [1978] valued improvements for rivers in a large western river basin. Sutherland and Walsh studied a large western lake. Vaughan and Russell [1982(a)] valued willingness to pay at fee fishing sites. Mitchell and Carson [1984] and Vaughan and Russell [1982(b)] both estimated aggregate willingness to pay for improvements in national water quality. In each case there is a serious lack of correspondence with the sample reaches.

The remaining three studies are Gramlich's study [1977] of the Charles River in Boston and two studies of the Monongahela River in western Pennsylvania by Smith and Desvousges [1986] and Smith et al. [1984]. The data and methodology in the Monongahela studies are much better than the Gramlich study, but all three studies included basic site characteristics and found that household differences (e.g., education and income) had little effect on willingness to pay for improved water quality. The travel cost estimates in Smith et al. include only user values, whereas the contingent valuation estimates of the other two studies include both use and non-use components of willingness to pay.

Both rivers involve warm-water recreation with good access. Activities in the Gramlich study included boating and shoreline recreation. The Monongahela studies included fishing and swimming as well. Both the Charles and Monongahela Rivers draw primarily on urban user populations. The recreation market in the Charles and Monongahela studies consisted of 716,200 households and 616,800 households, respectively. The mean respondent in the Charles study lived two miles from the site, and the most distant respondent lived eight miles from the site. The mean and maximum distances for the Monongahela studies were 15 miles and 40 miles, respectively.

The Monongahela studies estimated benefits for three changes in water quality (Table 3-I-1). Gramlich obtained a value only for a change from boatable to swimmable. For our purposes, we assume that swimmable corresponds to water that is suitable for game fishing.[1]

Table 3-I-1. Estimated Water Quality Benefits per Household in Three Studies (Annual Willingness To Pay -- $1984)

| River | Water Quality Change | | |
	Boat-Fish	Fish-Swim	Boat-Swim
Monongahela (CVM)	$25-40	$14-23	$40-64
Monongahela (TCM)	$8	$10	$18
Charles (CVM)	--	--	$74

There are two problems with transferring these estimates to the sample sites. First, the site characteristics of the Monongahela and Charles Rivers vary considerably from the sample reaches. Second, the water quality changes for the study rivers do not map neatly into the changes in the sample reaches. Even if we assume that fishable corresponds to rough fishing and swimmable corresponds to game fishing, we have no estimate for improvement from unusable to boatable.

Note

1. The water quality simulations provided data only on dissolved oxygen, which is an important factor for fishing potential. Swimming potential depends primarily on fecal coliform levels. We are therefore assuming that dissolved oxygen is an acceptable proxy for general water quality, including bacterial levels.

References

Brookshire, Daniel S., Larry S. Eubanks, and Charles F. Sorg. 1986. "Existence Values and Normative Economics: Implications for Valuing Water Resources." *Water Resources Research*, 22:1509-1518.

Gramlich, Frederick W. 1977. "The Demand for Clear Water: The Case of the Charles River." *National Tax Journal*, 30:183-94.

Mitchell, Robert C., and Richard T. Carson, 1984. *A Contingent Valuation Estimate of National Freshwater Benefits; Technical Report to the U.S. Environmental Protection Agency.* Resources for the Future, Inc., Washington, DC.

Naughton, Michael C., and William H. Desvousges. 1986. "Water Quality Benefits of Additional Pollution Control in the Pulp and Paper Industry." Report to the U.S. Environmental Protection Agency. Research Triangle Institute, Research Triangle Park, NC.

Naughton, Michael C., George R. Parsons, and William H. Desvousges. 1987. "Benefits Transfer: Conceptual Problems in Estimating Water Quality Benefits Using Existing Studies." Unpublished paper. Economics Department, San Diego State University, San Diego, CA.

Smith, V. Kerry, and William H. Desvousges. 1986. *Measuring Water Quality Benefits.* Kluwer-Nijhoff Publishing, Boston, MA.

Smith, V. Kerry, William H. Desvousges, and Ann Fisher. 1984. "A Comparison of Direct and Indirect Methods for Estimating

Environmental Benefits." Working Paper No. 83-W32. Vanderbilt University, Nashville, TN.

Sutherland, Ronald J., and Richard G. Walsh. 1985. "Effect of Distance on the Preservation Value of Water Quality." *Land Economics*, 62:282-91.

Vaughan, William J., and Clifford S. Russell. 1982(a). "Valuing a Fishing Day: An Application of a Systematic Varying Parameter Model." *Land Economics,* 58:45-63.

_____. 1982(b). "Freshwater Recreational Fishing: The National Benefits of Water Pollution Control." Report to the U.S. Environmental Protection Agency. Resources for the Future, Inc., Washington, DC.

Walsh, Richard G., Douglas A. Greenley, Robert A. Young, John R. McKean, and Anthony A. Prato. 1978. "Option Values, Preservation Values and Recreational Benefits of Improved Water Quality: A Case Study of the South Platte River Basin, Colorado." EPA-600/5-78-001. Fort Collins, CO.

Appendix 3-J

SENSITIVITY ANALYSIS FOR BENEFIT CALCULATIONS

Table 3-J-1. Alternative Benefit Calculations ($1984 10^6)

Weights Used in Calculations			
No Category Change		Category	Range of
< 25%	> 25%	Change	Benefits Estimates
0.0	0.0	1.0	$15.80 - $53.60
0.0	0.25	1.0	$16.00 - $54.40
0.25	0.5	1.0	$16.50 - $56.70
0.5	0.75	1.0	$17.00 - $59.10
1.0	1.0	1.0	$17.90 - $62.90

Chapter 4

AMBIENT APPROACH

The Clean Air Act establishes the national goal of achieving air quality that promotes public health and welfare and, in most cases, mandates ambient-based standards as the primary regulatory approach for reducing emissions. The Act requires meeting ambient standards regardless of the size of the population protected and compliance costs. Consequently, the relationship between benefits and costs is likely to vary across sites, but not nearly to the extent or to the degree that it would with technology-based standards.

An assessment of the efficiency of an ambient-based standard in reducing environmental risks requires linking the costs of implementing a specific regulatory program and the benefits that accrue from it. These linkages for the Clean Air Act regulation of total suspended particulate (TSP) emissions from the pulp and paper industry are illustrated in Figure 4-1. This figure shows that, when implemented, a regulatory program based on ambient-based standards will impose pollution control costs and reduce TSP emissions. These emission reductions will improve air quality in the vicinity of pulp or pulp and paper mills. An improvement in air quality will decrease human exposure to air pollution, which will diminish adverse health and welfare effects. Benefit estimates are based on monetary valuations of people's willingness-to-pay for these improvements in health and welfare.

Similar to the previous chapter, we estimate the real resource costs required for major segments of the pulp and paper industry to comply with an ambient-based standard and the resulting reduction in TSP emissions.[1] The costs and emission reductions result primarily from existing mills meeting emission limits specified in state-issued operating permits and from new mills meeting emission limits specified in New Source Performance Standards.

Decreasing TSP emissions reduces risks to human health and welfare. Health benefits include mortality as well as acute and chronic morbidity.

Figure 4-1. Framework for Benefit-Cost Analysis of the Clean Air Act
              Ambient-Based Standards

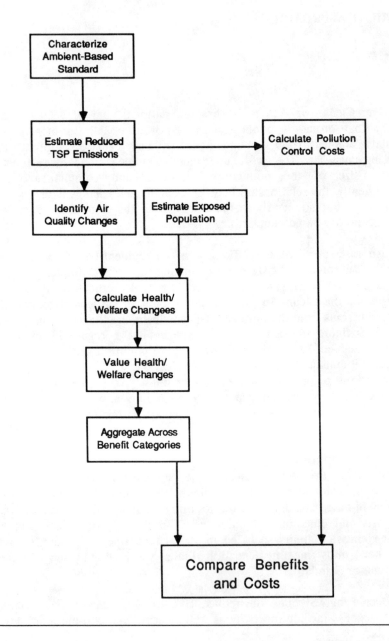

Welfare benefits include, among other things, reductions in household soiling and damages to materials, structures, and vegetation.

POLLUTION CONTROL COSTS AND EMISSIONS

We estimate the real resource pollution control costs and TSP emissions at three points in time - 1973, 1984, and 1994. Costs incurred before 1973 are not considered to be associated with the Clean Air Act because the regulatory requirements of the Act had not taken effect by 1973. (Actually, 1971 or 1972 might have been more appropriate, but 1973 was the first year national data were available on pollution control equipment in place and on emission discharges from individual mills.) Costs in 1984 are the costs for air pollution control under current regulations. Costs in 1994 are the future costs of pollution control, taking into account increased production but assuming the same New Source Performance Standards as in 1984. Other relevant types of cost and emission information are those associated with bringing existing mills into compliance with their emissions limits, those associated with 1984 production in the absence of federal regulations, and those associated with the introduction of technology-forcing regulations by 1994.

The main text of this chapter does not report our estimates of the costs for or emission reductions of the other significant criteria pollutant, sulfur dioxide (SO_2), from pulp and paper mills. These estimates are not part of the efficiency critique of the TSP ambient standard. We summarize our findings on SO_2 emissions and costs in Appendix 4-A.

Inventory -- 1973, 1984, and 1994

We identified 169 major pulp-producing mills operating in 1973 and 178 in 1984 and projected there would be 192 mills operating in 1994.[2] We could find sufficient data, however, for characterizing pollution control costs and emission loadings for only 150 mills in 1973, 157 mills in 1984 and 163 mills in 1994.

We describe the procedures used for establishing this inventory in Appendix 4-B and list the mills covered in Appendix B. The background data on these mill types are presented jointly for two of the three snapshot years, 1973 and 1984, and separately for 1994. The data categories for each year are: emission sources within mills, mean daily production, pollution control equipment in place, and compliance status.

Emissions Data

The National Emissions Data System provides TSP emission estimates for emission sources within each mill. Emission estimates for sources were checked against EPA's emission factors [EPA. 1985(b)]. Where they differed from these factors by plus or minus 50 percent, they were adjusted to be consistent with EPA's emission factors. Emissions from all sources were summed to give mill totals.

Simulated new process units at expanding and new plants are estimated to meet the emission limits specified in the New Source Performance Standards.

Cost Modeling

We constructed the Cost of Clean model to estimate source-specific costs for points in time and regulatory conditions [E.H. Pechan and Associates. 1987]. Figure 4-2 illustrates the basic structure of the model with its distinct algorithms for processes and boilers.

Baseline Costs. The Cost of Clean model computes the capital, operation and maintenance, recovery credits, and annual costs of the emission reduction processes that are in place in each of the three baseline years: 1973, 1984, and 1994. (See Appendix 4-C for a description of the specific cost functions.) Parallel to the computation of the emission reduction costs for each of the three baseline years, the model calculates the emission reductions and resulting emission levels and reports them in both pounds of TSP per ton of pulp production and tons of TSP in the given year.

The costs of TSP controls in 1973 represent an estimate of the costs of equipment installed at mills for the purposes of air pollution control. This is equipment over and above the level considered as economic recovery, and was installed as a result of state and local air pollution control regulations before any substantial requirements of the Clean Air Act.[3]

Particulate control costs for 1984 represent an estimate of the costs of equipment in place in 1984 above the equipment required for economic recovery. These costs can be considered to result from standards associated with the Clean Air Act and from regulations imposed by state and local agencies in the absence of Federal requirements.

Figure 4-2. Simplified Flowchart of the Cost of Clean Model -- TSP

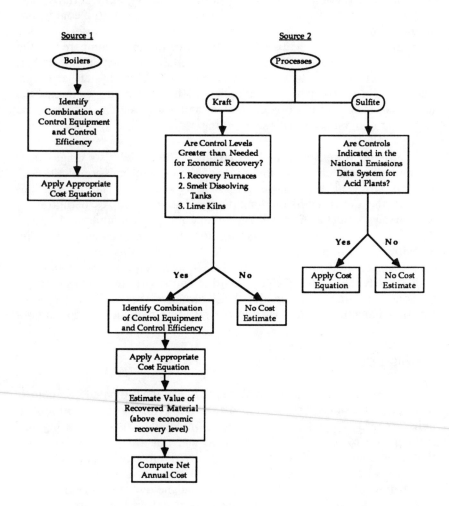

Particulate control costs for 1994 combine the 1984 costs with the additional costs of expanded and new mills meeting New Source Performance Standards. Costs and emissions are estimated accordingly.

Compliance Costs. The model also assesses the compliance costs for emission sources not in compliance with their state operating permits. The model does not, as it does for water dischargers, calculate the compliance status of emission sources. The task was too difficult because there are approximately 1,600 air emission points at the 157 mills compared to the one, or perhaps two, water effluent sources at each mill. Rather, compliance status is based on EPA's Compliance Data System and state records. If a source is out of compliance, then the model estimates the additional emission reduction equipment and associated costs to bring the source into compliance and the resulting reduction in emissions.

Costs of More Stringent Regulations. The model also assigns costs to more stringent emission limitations if the current emission limitations are less stringent than those proposed as future regulations. In this chapter, the future emission limitation is based on EPA's definition of Reasonably Available Control Technology for process units and boilers [FR. 1978, 1986]. If the emission limit for TSP for a given emission point is below EPA's definition, then the source is considered to be in conformance and is not analyzed further. If, however, the current emission limit exceeds the proposed limit, then the emission point is considered to be nonconforming. The model then calculates the costs of additional emission reduction technology and the resulting reduction in emissions.

Estimated Costs and Emissions

This section presents national summaries of the estimated TSP control costs and emissions for 1973, 1984, and 1994 for most pulp-producing mills. Cost estimates are presented for capital costs, operating and maintenance costs, credits for recovered materials, and net annual costs. Recovery credits are presented as a negative number to indicate that they represent value to the mill. The net annual costs are the sum of operation and maintenance costs, recovery credits, and the annualized capital costs. The emissions are the actual or calculated emissions for each year under different conditions for the period 1973 to 1994.

Some care is needed in interpreting the cost results, particularly those attributable to the Clean Air Act. To address this issue, we first calculated the actual costs in 1973 and 1984. We then estimated costs (and emissions) without the Clean Air Act by combining 1984 production levels with 1973 emission reduction processes beyond those associated with economic recovery of chemicals. The without-Clean Air Act costs were then subtracted from the with-Clean Air Act costs, where costs represent 1984 production with 1984 pollution control. This procedure eliminates those emission reduction costs already incurred before 1973, plus any increase in emission reduction costs due to expanded production. The costs attributable to the Clean Air Act in the 1973-1984 period then reflect only the real resource costs for the industry to comply with the national ambient air quality standard for TSP.

1973. The annual costs in 1973 for controlling TSP emissions from kraft, sulfite, and groundwood pulping and from associated boiler operations are listed in Table 4-1. The net annual costs were estimated to be approximately $45 million. The annual costs were those above the economic recovery level and were attributable to state and local rules. The industry emitted an estimated 350,000 tons of TSP in 1973.

1984 without the Clean Air Act. The annual costs in 1984 if the Clean Air Act requirements had not been implemented (but with state and local requirements) for controlling TSP would have been about $55 million. The 1984 air emissions without the Clean Air Act would have been about 400,000 tons of TSP (1.1 times 1973 emissions).

1984 with the Clean Air Act. The annual costs in 1984 with the Clean Air Act requirements for controlling TSP were approximately $100 million compared with an estimated $55 million without the Clean Air Act. The additional investment of $45 million in TSP emission controls resulted in a 190,000-ton decrease in TSP discharge instead of a 50,000-ton increase in TSP discharge between 1973 and 1984. The average cost per ton reduced was $150.

1984 with Compliance. The annual costs in 1984 with full compliance with the Clean Air Act requirements for controlling TSP could not be estimated for all 17 non-complying mills. We could find sufficient data for cost estimation at only seven mills. An incremental annual cost of $2.7 million would have reduced visible emission violations from power boilers at these seven mills. The incremental reduction in TSP would have been approximately 1,000 tons.

Table 4-1.   National Summary of TSP Pollution Control Costs and
                Emissions for Pulp Mills in This Study ($1984 10^6)

Scenario	Capital	O&M	Credits	Net Annual	Emissions (tons 10^3)
		Costs			
1973	$210	$30	($15)	$45	350
1984 w/o CAA	$235	$40	($15)	$55	400
1984 w/ CAA	$475	$55	($20)	$100	160
1984 w/ full compliance	$490	$60	($20)	$105	160
1994	$815	$70	($25)	$155	175
1994 w/ RACT	$1,080	$95	($35)	$205	115

1994. The annual costs in 1994 to meet 1984 existing and new source emission limits for TSP are projected to be approximately $155 million. The 1994 annual costs are 1.5 times the 1984 annual costs. In spite of these additional expenditures, TSP emissions from pulping and power boiler operations would increase by approximately 10 percent as a result of a 15 percent increase in capacity.

1994 with Reasonably Available Control Technology. The annual costs to comply with uniform Reasonably Available Control Technology for TSP are projected to be $205 million. The incremental annual costs of $50 million would result in a 60,000-ton reduction of the 175,000 tons of TSP emitted in 1984. This requirement would reduce emissions from process units and boilers at approximately 80 mills.

The costs and emissions estimates do not represent the total costs of pollution control or emission reductions on the part of the entire industry. However, they represent most of the costs and emissions based on comparisons with total industry capacity and emissions in 1973 and 1984. The mills covered in this analysis accounted for 97 percent of 1973 pulp production and 60 percent of 1973 emissions of TSP, and 94 percent of

1984 pulp production and 90 percent of the 1984 emissions of TSP [Medynski. 1973; Brandes. 1984; EPA. 1973, 1984(a)].

Cost Comparisons

We compared our real resource estimates with other real resource and financial cost estimates. The only real cost estimate is in "The Cost of Clean Air and Water Report to Congress 1984" [EPA. 1984(b)]. The only credible estimate of financial costs is the "Pollution Abatement Costs and Expenditures, 1984" [Bureau of the Census. 1986]. We present the results of the comparisons in Appendix 4-D. In general, the real resource cost estimates for TSP and TRS emission reductions from the Cost of Clean model are only 70 percent of the EPA cost estimates for annual costs.   They are lower because we used a lower capital recovery factor to annualize the capital costs and assumed higher credits for recovered materials, which lowers the operation and maintenance costs. The Cost of Clean model real resource cost estimates for TSP and TRS emission reductions are approximately equal to the financial cost estimates reported by the Bureau of the Census. The Cost of Clean model operation and maintenance costs are only 70 percent of the Census costs because we assumed higher credits for recovered materials.

AIR QUALITY IMPACTS

The benefits of improved air quality under the Clean Air Act must be measured by changes in specific air quality parameters. In general, such parameters should reflect the actual emissions discharged into the air and should affect human health and man-made and natural environments. In this study, we followed EPA convention in adopting TSP as the pollutant indicator for particulate matter. EPA used TSP as the pollutant indicator for particulate matter between 1971 and 1987. EPA recently replaced TSP as the indicator with one that includes only those particles with a diameter less than or equal to a nominal 10 micrometers (PM_{10}). The smaller particle sizes are more appropriate because they penetrate furthest in the respiratory tract and cause the greatest health risk.

Sample Sites

As stated earlier, we modeled the TSP emissions from approximately 157 mills for the year 1984. Unlike the water quality analysis, we did not need to limit the air quality analysis to only those areas where pulp mills were the only industrial source of TSP emissions. The air quality model allowed us to calculate the incremental change in ambient air quality concentrations due to controlling pulp mill emissions independently of other sources.

We did not model the changes in ambient air quality from reductions in TSP emissions for all 157 mills, however. Instead, we modeled only those mills that exhibited consistent emission and cost changes and showed at least a 100-ton difference between the with-and-without Clean Air Act scenarios. In addition, we limited our analysis to mills that were clustered reasonably near each other (53 counties) in order to minimize the costs of the modeling effort. As a result of these constraints, we limited our TSP benefit analysis to 60 mills located fairly evenly across the four pulp and paper-producing regions in the U.S. (Figure 4-3). These mills accounted for 40 percent of the TSP emissions from pulp mill operations in 1984.

Scenarios

Once the mills were selected, we assembled data for estimating TSP emissions for three scenarios. Scenario 1 portrays 1973 actual emissions from pulp mills based on National Emissions Data System data. It approximates conditions before the Clean Air Act by applying 1973 emission factors to 1973 production. Scenario 2 portrays 1984 hypothetical emissions from pulp mills. It approximates emissions without the Clean Air Act by applying 1973 emission limitations to 1984 production. Scenario 3 portrays 1984 actual emissions from pulp mills in compliance with their operating permits based on National Emissions Data System data. It approximates the Clean Air Act's emission limitations applied to 1984 production. Emission loadings for the 60 mills for the three scenarios are presented in Appendix 4-E.

Air Quality Modeling

Previous benefit studies have used a variety of methods to calculate the magnitude of air quality improvements that would result from air pollution controls. The usual approach, linear rollback modeling, assumes that reductions in pollutant concentrations are strictly proportional to reductions in emissions. Concentration reductions are estimated by reducing the monitored ambient concentrations, above the estimated background concentration, by an amount equal to the proportional

Figure 4-3. Map of Kraft Pulp Mills Included in the Air (TSP) Analysis

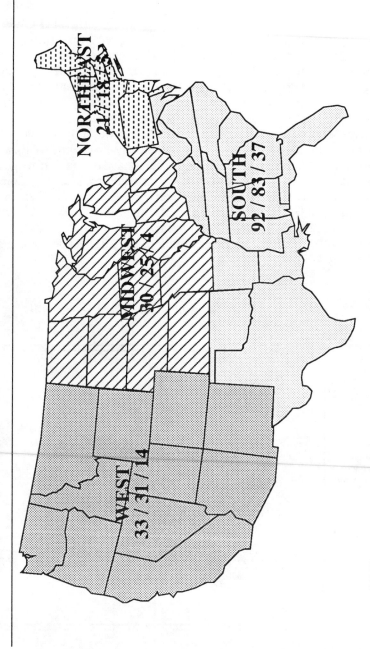

LEGEND: No. of pulp producing mills / no. of mills with cost estimates / no. of mills in benefit-cost analysis

reduction in emissions. A major weakness in linear rollback modeling is that it essentially ignores source-receptor relationships because it assumes that sources of pollutants affect all receptors equally.

A more resource-intensive alternative is to use air quality dispersion models to predict concentration changes in a given area attributable to emission reductions by specific sources. This approach accounts for the source-receptor relationships for those sources modeled and provides a more accurate estimate of the ambient concentration reductions due to emission reductions for the modeled sources. Historically, such air quality dispersion models as the Industrial Source Complex model are used sparingly because of the expense. EPA has recently developed an inexpensive alternative, however, the Model City Program (MCP) model [Maxwell and Manuel. 1985]. This model is a data base management system that mimics the performance of the Industrial Source Complex dispersion model. It provides a relatively inexpensive method of estimating the ground-level impact for large numbers of sources across the United States. We selected this new model because it is more accurate than the linear rollback model. A brief description of the MCP model is presented in Appendix 4-F.

Figure 4-4.   Pollution Dispersion Around the Union Camp Mill
              (annual average)

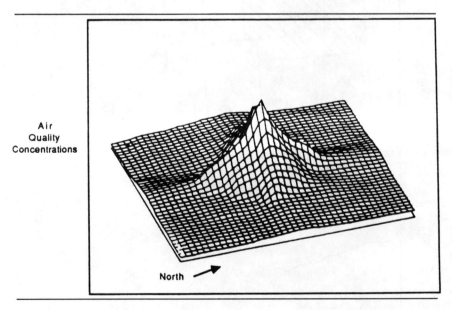

The MCP model portrays annual arithmetic mean concentrations in a region (39 km x 39 km square) with values at 1-km grid points. Output from the MCP model indicates significant pulp mill-related pollution within a radius of 17 km (10 miles) around the mills and negligible pollution outside this area. Figure 4-4 illustrates the annual mean dispersion of pollutants around the Union Camp mill located in Alabama.

For all three scenarios, we added the MCP concentration values to the monitored 1984 county average values to estimate ambient concentrations. We computed summary air quality values for five rings around the mill based on the average value of all MCP grid points within each annular area. The annular areas selected were 0-1.0, 1.1-2.0, 2.1-3.0, 3.1-5.0, and 5.1-10.0 miles.

Summary air quality profiles for four of the 60 mills are presented in Table 4-2 to show how we aggregated the model's outputs and interpreted them for air quality impact assessments. For example, TSP emissions from Mill 704 would have increased from 3,100 to 6,800 tons between 1973 and 1984 due to increased production. They actually decreased to 1,100 tons, however, because of the Clean Air Act. The hypothetical 1984 emissions without the Clean Air Act would have caused TSP concentrations in excess of 50 μg/m^3. Neither the 1973 nor the 1984 emissions resulted in TSP concentrations above 50 μg/m^3, however.

Although the change in emissions from the 1984 levels without the Clean Air Act to the 1984 levels for Mills 704 and 714 was the same (5,700 tons), the difference in TSP concentrations is greater for Mill 714 than for Mill 704. Also, even though Mill 714 had lower total emissions in 1984 without the Clean Air Act compared to Mill 704, its emissions resulted in higher contributions to TSP ambient concentrations. Whereas the linear rollback model would not have predicted these differences, the MCP model does predict these differences because it accounts for stack height differences at each plant and meteorological conditions.

Estimated Air Quality Improvements

The summary results of the air quality simulations without (Scenario 2) and with (Scenario 3) the Clean Air Act emission limits for 60 mills showed varied outcomes (Figure 4-5). At the outset, we should point out that these results must be interpreted differently from the water quality results because, as will be described in the health effects section, there are

Table 4-2.   Annual Average Concentration Values for
Four Pulp-Producing Mills

Scenario	Emissions (tons)	TSP Concentrations ($\mu g/m^3$) For Five Annular Areas (mi)					Average
		0-1	1-2	2-3	3-5	5-10	
Mill 704							
1973	3,100	49.2	48.8	48.4	48.0	47.5	48.4
1984 w/o CAA	6,800	51.9	51.0	50.1	49.2	48.1	50.1
1984 w/ CAA	1,100	47.8	47.6	47.5	47.3	47.2	47.3
Mill 714							
1973	5,100	52.7	50.8	49.6	47.9	46.4	49.4
1984 w/o CAA	6,400	54.6	52.2	50.6	48.7	46.8	50.6
1984 w/ CAA	700	46.1	45.8	45.6	45.4	45.2	45.6
Mill 735							
1973	5,000	84.4	61.6	54.3	45.3	42.5	47.6
1984 w/o CAA	6,500	97.4	67.6	58.0	46.3	42.7	62.4
1984 w/ CAA	900	50.0	45.7	44.3	42.6	42.1	44.9
Mill 835							
1973	7,000	72.3	69.3	67.4	65.2	63.4	67.5
1984 w/o CAA	6,900	72.1	69.1	67.3	65.2	63.3	67.4
1984 w/ CAA	1,800	64.6	63.8	63.4	62.8	62.3	63.4

primary standard for TSP of 75 $\mu g/m^3$ (annual average) was set as the threshold for health effects (albeit in 1971), the health effects studies used in this study suggest that there is no threshold level for mortality effects and that there is a 40 $\mu g/m^3$ threshold level for morbidity effects. As a result, there would be health effects and thus monetary benefits associated with any change in ambient concentration even though a mill remains in the same category (any one of the six microgram ranges) in Figure 4-5.

With that perspective, we point out that emissions without the Clean Air Act from nine of the 60 mills would have exceeded the primary health standard, and 58 of the 60 mills would have exceeded the morbidity threshold. With the Clean Air Act, only four of the 60 mills exceed the primary health standard, whereas 58 still exceed the morbidity threshold.

In addition, the number of mills with emission concentrations in the range of 60-70 μg/m^3 falls from 11 to 5.

Figure 4-5.   Air Quality Around 60 Mills With and Without
the Clean Air Act

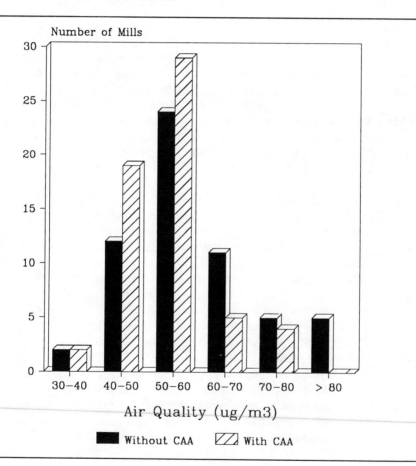

The resulting reductions in the estimated annual average TSP concentrations, using the maximum annual average concentration for each mill for comparison, ranged from 0 μg/m^3 to 122 μg/m^3. The mills with the largest reductions in ground-level TSP concentrations were not usually the mills with the largest emission reductions. Appendix 4-G compares the 20 mills with the largest reductions in concentrations with the 20 mills with the largest emission reductions. In general, stack height appears to be more important in determining changes in ambient concentrations that emission reductions, which reinforces the importance of using the MCP model.

Validation of MCP Model

The accuracy of the ground-level pollutant concentration estimates are critical in any dispersion modeling exercise. We could not use monitoring data as we did to verify the water quality modeling. There were no appropriately located air quality monitors around the 60 sample sites. Even if there were monitors, verification with ambient data would have been impossible because of the difficulty of accounting for all stationary sources and fugitive emissions. Instead, we followed the traditional verification approach, which compares the results of two different models.

We compared the results of the MCP model with the results of EPA's established Industrial Source Complex Long Term dispersion model in two different settings. The first comparison, a product of developing the MCP model, used the MCP and Industrial Source Complex models to estimate TSP concentrations in and around Baltimore, Maryland [Maxwell and Mahn. 1988]. In general, the annual average concentrations predicted by the two models are similar for rural and suburban areas. However, the MCP ambient estimates exhibit a downward bias of about one $\mu g/m^3$. The second comparison, a product of this study, used the MCP and Industrial Source Complex models to estimate TSP concentrations for 14 pulp and paper mills in this study [Maxwell et al. 1987]. Overall, the MCP estimates are in reasonable agreement with the Industrial Source Complex results, given the difference in receptor locations. However, as in the Baltimore comparison, the Industrial Source Complex maximum concentrations are larger than the MCP maximum concentrations. More details about the comparisons are in Appendix 4-H.

MONETARY BENEFITS

Society values improvements in air quality because the improvements reduce adverse health and welfare effects. Estimation of societal values in monetary terms for these improvements requires assessment of the changes in exposure to air pollution, calculation of the physical changes in potential adverse health and welfare effects, and valuation of the physical changes in monetary terms [Freeman. 1979].

Exposure

Exposure is defined as the product of the ambient concentration of a pollutant and the exposed population. Since both variables can vary spatially and temporally, obtaining precise measures of human exposure is difficult. Therefore, we introduced simplifying assumptions to make the quantitative measurement of exposure more tractable.

The Graphical Exposure Modeling System generated estimates of the population residing within a circular area surrounding each mill and extending in radius by 10 miles. This system uses 1980 Census population data that were adjusted to 1984 levels using area growth rates based on historical data. The data are disaggregated into a 1-mile-radius circular region centered on the mill, with annular rings at 2-, 3-, 5-, and 10-mile radii.

Exposure is characterized by assuming that the populations living in each of these rings experience a uniform level of air quality equal to the mean value of each set of 1-km grid points within each ring. The total exposures for four of the 60 modeled mills are presented in Table 4-3 to show the output of the model and how we interpreted them for benefit-cost analysis. These four mills are the same as those examined earlier. Total exposure is simply people times concentration. For example, TSP concentrations around Mill 704 decreased on average by 0.9 $\mu g/m^3$ as a result of the Clean Air Act. Total exposure decreased by 30,000 units. TSP concentrations around Mill 714 decreased in the one ring (5-10 miles) where people live on average by 1.2 $\mu g/m^3$ as a result of the Clean Air Act. Total exposure dropped by 20,000 units. The change in total exposure is relatively small, even though the average change across all rings is quite large (3.4 $\mu g/m^3$) because people live only where the exposure change is the smallest.

The summary results of the exposure changes for the 60 evaluated mills show varied outcomes (Figure 4-6). First, note that the exposure changes for the 60 evaluated mills are similar to the exposure changes for all 111 mills (including the 60) with modeled air quality changes. Second, approximately one-half of the exposure changes falls into the 10,000 - 100,000 range and the mean change for the 60 mills is 60,000 units. Most

of the remaining changes are in the smallest range of 1-10,000, indicating
both minor differences in TSP concentrations and few people exposed.

Table 4-3. Total Exposures for Four Pulp Mills

Scenario	Average Air Quality in Populated Rings ($\mu g/m^3$)	People Within 10 Mile Radius (10^3)	People x Air Quality (10^3)
Mill 704			
1984 w/o CAA	48.1	33,700	1,620
1984 w/ CAA	47.2	33,700	1,590
Difference	0.9	-	30
Mill 714			
1984 w/o CAA	48.6	10,500	490
1984 w/ CAA	45.2	10,500	470
Difference	3.4	-	20
Mill 735			
1984 w/o CAA	45.1	124,700	5,620
1984 w/ CAA	42.4	124,700	5,290
Difference	2.7	-	330
Mill 835			
1984 w/o CAA	66.5	21,500	1,430
1984 w/ CAA	63.3	21,500	1,360
Difference	3.2	-	70

Health and Welfare Effects

Potentially relevant health and welfare effects of TSP reductions are
listed below:

Effects Categories Potentially Relevant for a TSP Benefit Analysis

Health Effects
• Mortality
• Acute Morbidity
• Chronic Morbidity

Acidic Deposition
• Aquatic & Terrestrial Life
• Crops and Forests
• Materials

Soiling & Material Damage
- Residences
- Commerce and Industry
- Institutions
- Government

Climate Effects
- Temperature
- Precipitation

Visibility Effects
- Regional Haze
- Plume Blight

Non-Use Benefits
- Bequest Value
- Existence Value
- Option Value

Although the literature contains many studies that show an association between TSP exposure and various health effects, not all of these studies are useful for benefit analysis purposes [Mathtech. 1988]. At least two conditions must be met if the health effects are to be used in the benefit analysis: (1) the TSP-health effect association should be stated in the form of a concentration-response function, and (2) the health effect should be amenable to valuation.

These criteria limit the selection of health models. For this effort we drew on two studies: (1) Schwartz and Marcus [1990] -- a study that examines the relationship between British Smoke and daily mortality in London, and (2) Ostro [1987] -- a study that uses regression techniques to estimate the relationship between short-term exposure to fine particles and acute illness, measured in terms of respiratory-related reduced-activity days. We considered the Ferris model [Ferris et al. 1973, 1976] of chronic morbidity, but eliminated it because it is only relevant for annual average TSP concentrations greater than 130 $\mu g/m^3$, which concentration was never estimated by the MCP model.

For welfare effects, we drew on only one study: Mathtech household soiling [1982, 1986] -- a study that examines the relationship between TSP and the demand of households for various goods and services. It is not possible to evaluate benefits to other sectors of the economy or to evaluate welfare effects other than reduced soiling. Because the coverage of effects in both the health and welfare areas is incomplete, the benefit estimates in this report understate actual benefits. However, this selection of models is consistent with many previous benefit analyses conducted for EPA.

Mortality. Schwartz and Marcus reanalyzed the London data used in earlier studies on the relationship between daily variations in mortality rates and changes in particulate matter, measured as British Smoke and SO_2 concentrations. Their report presents results for several different specifications of the concentration-response functions and uses TSP, rather than the British Smoke used in earlier studies, as the pollutant parameter for particulate matter. They found mortality health effects at exposures as low as background levels. For a 1-μg/m^3 change in TSP and a population of one million, their mid-point estimate of the number of expected mortalities is one and one-half. Their lower-bound estimate is approximately one expected mortality, and their upper-bound estimate is approximately two expected mortalities. Appendix 4-I describes the derivation of these estimates from their study.

Figure 4-6. Comparison of Exposure Change Profiles

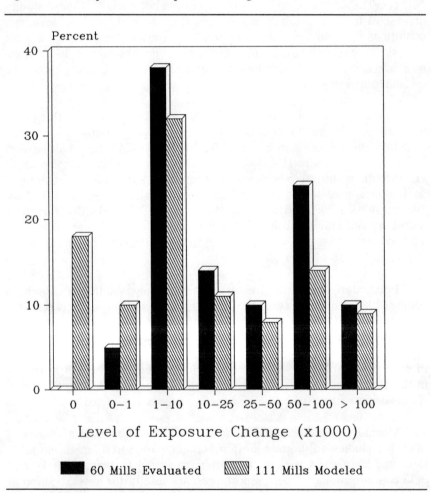

Morbidity. The best available study of the relationship between morbidity and TSP concentrations is Ostro [1987]. Ostro uses morbidity data from the Health Interview Surveys for 1976-1981. He combines these data with various demographic measures and data on fine particle concentrations developed from airport visibility data. Fine particle data are thought to be the appropriate relevant measure of pollution for TSP-related morbidity effects. A Poisson distribution is used to model several health effects including work-loss days, reduced-activity days, and respiratory-related reduced-activity days. Ostro also used a fixed-effects specification to control for unobserved differences in the cross-section data.

Because of the extensive evidence linking particulate matter exposure to respiratory illness, we limited the morbidity health effects that we estimated to respiratory illness. We used Ostro's respiratory-related activity day concentration function that indicated potential effects as low as 40 $\mu g/m^3$. For a one $\mu g/m^3$ decrement in mean annual TSP and a population of one million, the mid-point estimate of the reduced number of work-loss days is 4,500 over one year, and the mid-point estimate of the diminished number of respiratory-related reduced-activity days is 27,000. The lower-bound estimates are approximately 3,000 work-loss days and 19,000 restricted-activity days, and the upper-bound estimates are approximately 9,950 work-loss days and 60,000 respiratory-related restricted-activity days. Appendix 4-I describes the derivation of these effects estimated from the Ostro study.

Valuation of Benefits

The logical common unit of measure for various health and welfare benefits is dollars, especially because the costs of controlling emissions are stated in dollars.

Mortality Values. The value of mortality benefits is the value of small reductions in the risk of death. Note that the value is not the willingness of any particular person to pay to prevent certain death. The method used to value risk reductions relies on studies that examine the wage differential required by workers to accept risky jobs [EPA. 1983]. The results of these studies indicate that workers require an extra $100 to $700 annually to accept an additional mortality risk of 1 x 10^{-4}. In more familiar terms, this is equivalent to a value of $1 to $7 million per statistical life saved. In this analysis, $1.7 million (1984 dollars) is used as a point estimate, with a range of $0.4 to $3.2 million. (The high-end valuation was halved from the EPA guidelines [1983] in order to form a more reasonable estimate.)

There are limitations associated with this approach to mortality risk valuation. Workers in hazardous jobs may have different risk preferences than the general population. If the workers from which the wage premium data were taken are relatively greater risk takers, then the risk compensation required to avoid a unit change in risk would be understated for the general population.

Morbidity Values. The correct measure of the value individuals place on morbidity effects is their willingness-to-pay to avoid the effect. Since willingness-to-pay is difficult to measure, most attention to value morbidity relies on the cost of illness. This is the approach adopted for the present study.

EPA's guidelines value three components of the cost of illness [EPA. 1985(a)]. The first measure is the value of a work-loss day. A work-loss day is valued at the wage rate (i.e., the opportunity cost of time). The second element of cost is the value of a reduced-activity day. This component is valued at one-half of the average daily wage. The third component of cost is the direct medical expenditure associated with morbidity. Only acute respiratory expenditures of $70 per capita are considered because the morbidity effects examined in this analysis are limited to acute respiratory conditions. In general, the measures of cost of illness underestimate willingness-to-pay because they omit willingness-to-pay for reduced pain and suffering experienced by affected individuals.

Household Soiling and Materials Damage. The hypothesis of the Mathtech model is that ambient TSP affects the demand for some market goods and that this impact can be interpreted as a measure of willingness-to-pay [Mathtech. 1982, 1986]. The model consists of a two-stage budgeting process. In the first stage, consumers allocate their income among broad groups of utility generating commodities such as food, clothing, and transportation. In the second stage, the broad group's budget (e.g., food) is further allocated among specific items. For example, one argument in the utility function might be cleanliness. The price and degree of cleanliness can be derived from observed market purchases of goods and services that contribute to the production of cleanliness. As air quality improves, an individual would purchase fewer goods that produce cleanliness in order to maintain the same level of cleanliness. This saving is the source of economic benefits.

Estimated Benefits

Table 4-4 combines the data on exposure, concentration-response functions, and values to obtain monetary estimates of benefits for all mills examined and to compare health and welfare benefits to pollution reduction costs. The three benefit estimates reflect possible combinations of concentration-response functions and monetary values to give low, mid-point, and high benefit estimates. The cost estimates are the differences in costs between the with and without Clean Air Act scenarios.

On a case-by-case basis, the number of mills that meet the benefit-cost criterion varies with the assumption about benefit values. Whereas at the lower-bound value only four mills meet the criterion, at the higher-bound value 33 mills meet it, and at the mid-point value 20 mills meet it. Whether a mill meets the criterion is a function of emissions, resulting ambient concentrations, and population exposed.

Possible Biases in the Benefit Calculations

Contrary to the water quality analyses, there are not obvious sources of bias in our reported benefit calculations (other than the omission of some benefit categories). One potential bias is that the population around the 60 evaluated mills is not the same as the population around the 157 study mills for which we had population data. Upon examination of the population distributions in Figure 4-7, however, we could not see any reason to believe that the population in our sample differed from the study population. If anything, the evaluated mills on average have slightly larger populations around them than the sample mills.

FINDINGS

The Clean Air Act ambient-based standard for TSP, as applied to the pulp and paper industry, met the net benefits criterion based on the 60 mills in our sample. As shown in Table 4-5, the conservatively estimated benefits exceed the costs based on the mid-point and the upper-bound benefit estimates. In addition, assuming that the mid-point benefit estimate best characterizes the environmental results, there are net benefits at 35 percent of the mills.

Table 4-4. Estimated Benefits and Costs of the Clean Air Act
for 60 Pulp Mills ($1984 10^3)

Mill	Change in Air Quality ($\mu g/m^3$)	Annual Benefits			Annual Costs
		Min	Mid	Max	
700	0.3	$10	$40	$90	$830
702	0.1	0	0	0	140
703	0.1	0	20	50	200
704	2.6	20	170	370	470
705	0.8	0	30	60	590
708	4.9	10	90	190	160
712	0.8	0	10	10	260
713	0.7	10	100	220	1,910
716	1.5	10	50	100	410
718	0.7	10	50	120	230
720	2.8	80	630	1,380	160
723	0.4	0	30	60	70
724	3.6	220	1,770	3,890	30
725	1.8	10	70	150	400
729	0.3	20	160	350	540
730	0.1	10	40	90	780
731	5.2	20	140	300	170
735	7.6	220	1,760	3,840	510
736	7.8	220	1,730	3,800	350
737	0.2	10	70	140	1,200
739	3.2	70	500	1,180	520
741	1.7	90	740	1,630	330
742	0.2	0	20	50	450
743	0.8	110	900	2,020	1,050
744	0.2	10	80	180	470
747	1.5	20	160	360	390
748	2.8	10	40	80	550
751	1.1	50	410	900	330
754	2.4	200	1,780	3,900	760
759	0.4	0	20	40	40
761	0.9	10	130	280	120
764	0.8	10	80	170	60
766	1.1	100	720	1,560	780
770	3.8	30	240	530	510
772	3.4	40	330	710	690
776	7.5	0	20	50	40
786	0.5	10	80	170	20
788	2.1	90	730	1,610	870
791	3.2	10	100	230	90
795	1.3	110	940	2,070	400
798	0.1	0	0	10	10
804	5.6	50	440	960	140
808	0.1	0	30	70	20
814	0.1	0	0	10	430
817	0.4	10	80	170	290
820	0.3	50	430	930	530
822	0.7	40	320	690	700
823	0.4	0	30	70	70
827	0.5	10	50	110	160
828	0.3	10	60	120	190
835	4.1	130	1,070	2,380	750
837	0.3	30	250	600	1,240
841	0.5	10	60	130	270
855	2.4	200	1,750	3,870	330
858	5.1	250	2,000	4,320	10
860	0.9	80	660	1,450	80
864	1.9	10	70	160	220
868	5.1	10	70	150	280
880	0.7	80	680	1,490	500
881	5.1	240	2,050	4,470	70
Totals		$3,060	$25,080	$55,090	$24,170
Average	1.9	$50	$420	$920	$400

Figure 4-7. Comparison of Population Profiles

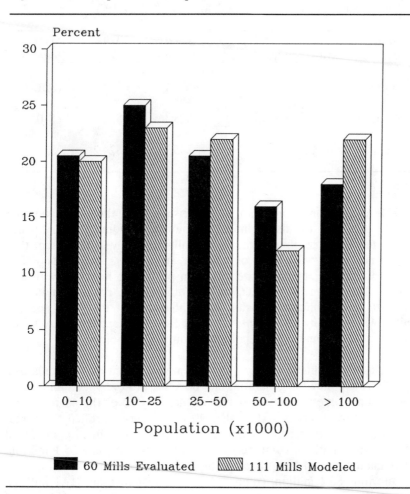

Earlier, we estimated the total costs of the Clean Air Act ambient-based standard for controlling TSP emissions at $46.8 million for 157 mills. The average annual Clean Air Act pollution control cost is approximately $300,000 per mill; the average annual pollution control cost for the 60 sample mills is $400,000 per mill. Assuming that the 60 sample mills are representative of the 157 mills, the benefits of TSP pollution control attributable to the Clean Air Act are most likely equal to and possibly two times greater than the costs of pollution control attributable to the Clean Air Act.

Table 4-5.   Benefits and Costs Attributable to the Clean Air Act
Ambient-Based Standard for TSP ($1984 10^6)

Valuation Scheme	Mills with B>C / Mills Analyzed	Total Benefits	Total Costs	Net Benefits
Minimum	4 / 60	$ 3.1	$24.2	($21.1)
Mid-Point	20 / 60	$25.1	$24.2	$ 0.9
Maximum	33 / 60	$55.1	$24.2	$30.9

This conclusion should not be surprising. On the one hand, the Clean Air Act requires existing industrial sources, including pulp and paper mills, to reduce emissions of criteria air pollutants only to the extent that they do not cause violations of ambient air quality standards.   The ambient air quality standards are based on protecting human health and welfare.   The Clean Water Act, on the other hand, requires industrial sources to meet uniform emission limits that do not take into account ambient water quality conditions.   Instead, these effluent limits are based on the availability and cost of pollution control technology.

A closer examination reveals that even an ambient-based standard is not a discrete policy for generating net benefits, however. Assuming that the mid-point benefit estimate is reasonable, only 20 of the 60 mills met the benefit-cost criterion.   The other 40 mills over-controlled their TSP emissions. The estimated annual benefits of pollution reduction at the 20 mills are $18.7 million, compared with annual costs of only $5.1 million. The estimated annual benefits of pollution reduction at the remaining 40 mills are only $6.4 million, compared with annual costs of $19.1 million. Thus, pollution reduction at 35 percent of the mills included in the benefit-cost analysis achieved 75 percent of the total benefits for only 20 percent of the total costs.

The higher-than-average cost of pollution reduction at the 40 over-controlled mills ($465,000 compared to $300,000 for all 157 mills in the study) makes generalizing about excess emission reduction difficult. There is no reason to doubt, however, that this proportion of over-controlled mills applies to all 157 mills. Probably 50 to 60 percent of the mills over-controlled their emissions when their costs are compared to the potential benefits. This finding suggests that ambient standards are crude policy instruments for achieving environmental improvements that result in net benefits to society. Moreover, as will be mentioned in Chapter 9, the overall favorable findings about TSP ambient standards do not necessarily apply to all ambient air quality standards.

Notes

1. Kraft (sulfate) and sulfite pulping are the two major pulping processes generating TSP emissions in the pulp and paper industry. The major sources of TSP emissions within a kraft mill are recovery furnaces, digesters, multiple-effect evaporators, smelt dissolving tanks, lime kilns, and power boilers. The major emission sources within sulfite mills are acid plants and power boilers. Other sources, such as blow tanks at sulfite mills and other mill types such as neutral sulfite semichemical and groundwood or thermomechanical, are not considered major sources of TSP. Groundwood mills as a subcategory discharge substantially more SO_2, NO_x, and CO than do sulfite mills because groundwood production in the United States is substantially larger than sulfite production. Most of these discharges are from the power boilers.

2. The basis for the 1994 projection of capacity is explained in Appendix 4-B.

3. Pulp mills, especially kraft mills, are unusual air pollution sources because significant reductions in emissions occur as a result of the economic recovery of valuable materials. In the process of producing kraft pulp from wood, mills use chemicals and heat to dissolve lignin from wood and then recover both the inorganic cooking chemicals and the heat content of the black liquor (a combination of spent cooking liquor and dissolved lignin), which is separated from the cooked pulp. An economically necessary step, recovery is accomplished by concentrating the liquor in evaporators to about 55 to 70 percent solids and then feeding the liquor to a recovery furnace, where combustion and chemical recovery take place.

As defined by Arthur D. Little [1977], economic recovery levels in all three snapshot years are as follows: recovery furnace - 95 percent; smelt dissolving tank - 85 percent; lime kiln - 95 percent. For the purposes of this study, particulate control at or below these levels is not considered air pollution control, and the application of the control equipment is not considered a cost associated with the Clean Air Act or state air pollution regulations.

References

Arthur D. Little. 1977. "Economic Impacts of Pulp and Paper Industry Compliance with Environmental Regulations." EPA-230/3-76-014-1. Cambridge, MA.

Brandes, Debra A. (ed.). 1984. *1985 Post's Pulp and Paper Directory*. Miller Freeman Publications, San Francisco, CA.

E.H. Pechan and Associates. 1987. "Air Pollution Control Costs For Pulp and Paper Mills." Draft report to the Office of Policy, Planning, and Evaluation, U.S. Environmental Protection Agency. Springfield, VA.

Federal Register (FR). 1978. "Standards of Performance for Stationary Sources: Kraft Pulp Mills." 43:7568-7596.

_____. 1986. "Standards of Performance for New Stationary Sources; Industrial-Commercial-Institutional Steam Generating Units." 51:42796-42797.

Ferris, B.G., Jr., H. Chen, S. Pules, and R.L.H. Murphy, Jr. 1976. "Chronic Nonspecific Respiratory Disease in Berlin, New Hampshire, 1967-1973: A Further Follow-up Study." *American Review of Respiratory Disease*, 113:475-485.

Ferris, B.G., Jr., I. Higgins, M.W. Higgins, and J.M. Peters. 1973. "Chronic Nonspecific Respiratory Disease in Berlin, New Hampshire, 1961-1967: A Further Follow-up Study." *American Review of Respiratory Disease*, 107:110-122.

Freeman, A. Myrick III. 1979. *The Benefits of Environmental Improvements*. Johns Hopkins University Press, Baltimore, MD.

Mathtech, Inc. 1982, 1986. "Benefits Analysis of Alternative Secondary National Ambient Air Quality Standards for Sulfur Dioxide and Total Suspended Particulates, Volume II." Report to the U.S. Environmental Protection Agency. Princeton, NJ.

_____. 1988. "Preliminary Benefit Analysis of Air Pollution Control Regulations For Selected Pulp and Paper Mills." Draft report to the Office of Policy, Planning, and Evaluation, U.S. Environmental Protection Agency. Princeton, NJ.

Maxwell, Christopher D., and Ernest H. Manuel. 1985. "Description of the Model City Program Methodology and Development," Versar, Inc., Columbia, MD.

Maxwell, Christopher D., Ducan McHale, Sally A. Mahn, and Keith Timbre. 1987. "Sensitivity Analysis of the Model City Program." Report to the U.S. Environmental Protection Agency. Versar, Inc., Columbia, MD.

Maxwell, Christopher D., and Sally A. Mahn. 1988. "Assessment of the Performance of the Model City Program." Report to the U.S. Environmental Protection Agency. Versar, Inc., Columbia, MD.

Medynski, Ann L. (ed.). 1973. *Post's 1974 Pulp and Paper Directory*. Miller Freeman Publications, San Francisco, CA.

Ostro, Bart D. 1987. "Air Pollution and Morbidity Revisited: A Specification Test." *Journal of Environmental Economics and Management*, 14:87-98.

Schwartz, Joel, and Allen Marcus. 1990. "Mortality and Air Pollution in London: A Time Series Analysis." *American Journal of Epidemiology*, 131:185-194.

U.S. Bureau of the Census. 1986. "Pollution Abatement Costs and Expenditures." MA-200(84)-1. U.S. Government Printing Office, Washington, DC.

U.S. Environmental Protection Agency (EPA). 1973. "1973 National Emissions Report." Office of Air Quality Planning and Standards, Research Triangle Park, NC.

_____. 1983. "Valuing Reductions in Risks: A Review of the Empirical Estimates." EPA-230-05-83-003. Washington, DC.

_____. 1984(a). "1984 National Emissions Report." Office of Air Quality Planning and Standards. Research Triangle Park, NC.

_____. 1984(b). "The Cost of Clean Air and Water Report to Congress 1984." EPA-230-05-84-008. Office of Policy Analysis, Washington, DC.

_____. 1985(a). "Estimates of Willingness to Pay for Pollution - Induced Changes in Morbidity: A Critique for Benefit Cost Analysis of Pollution Regulation." EPA-230-07-85-009. Washington, DC.

_____. 1985(b). "Compilation of Air Pollutant Emission Factors: Stationary Point and Area Source." AP-42, Fourth Edition. Office of Air Quality Planning and Standards. Research Triangle Park, NC.

Appendix 4-A

SO₂ EMISSIONS AND COSTS

The primary sources of SO_2 emissions from mills producing pulp are power boilers that are associated with all three major pulping processes. In addition, the blow pits and acid plants at sulfite pulping mills emit SO_2.

As in all industries, pollution control of SO_2 emissions from boilers is accomplished by: (1) switching to low sulfur coal, (2) switching to alternative fuels, and (3) adding scrubbing equipment to remove SO_2.

The pulp and paper industry took steps to reduce SO_2 emissions between 1973 and 1984. The industry increased its use of wood and wood waste and improved chemical recovery with steam production. In addition, the industry used a large number of wet scrubbers, primarily for particulate control. In order to prevent corrosion and remove SO_2, they added caustic materials such as soda ash (both purchased and generated in the process) to the scrubbing system. Moderate levels of SO_2 removal were common at mills using this approach.

From a control cost estimation standpoint, we found it difficult to separate the costs of TSP and SO_2 controls. Many scrubbers are used on boilers capable of firing wood to reduce TSP emissions. We reviewed EPA's National Emissions Data System to identify any SO_2-only control devices. There were 12 SO_2 scrubbers in the industry as of 1985. Three are lime spray dryers, and the remainder are various wet scrubbing options. We could not find any boilers where wet scrubbers existed along with a purely TSP control device, such as a precipitator or fabric filter. We assumed that the costs of these systems were captured in the TSP costs. Consequently, the only costs for SO_2 emission reductions are the plant-specific data for the three spray dryers.

An alternative to scrubbing equipment for SO_2 removal is fuel switching. Our analysis examined boiler fuel use by fuel type and sulfur content for each plant that had records in both the 1973 and 1984 National Emissions Data System. We looked at only coal and oil fuel use and assigned prices for different ranges of sulfur values for each fuel (the lower the sulfur content, the higher the price). The result of the matching and calculation showed only one plant with a positive fuel switching cost from 1973 to 1984. On the basis of available data, we could not assign any costs to fuel switching for the period 1973 to 1984.

Table 4-A-1 summarizes the results of our analysis. The annual costs of SO_2 pollution control are trivial compared to those for TSP and TRS - - and the emission reduction is small, less than 10 percent, between 1973 and 1984.

Table 4-A-1. National Summary of SO_2 Pollution Control Costs
and Emissions for Pulp Mills in This Study ($1984 10^6)

	Costs			Emissions[a]
Scenario	Capital	O&M	Annual	(tons 10^3)
1973	$0	$0	$0	780
1984 w/o CAA	$0	$0	$0	900
1984 w/ CAA	$10	$2	$3	710
1994	$10	$2	$3	970

a. Note that there is considerable uncertainty in the emission estimates listed. As an example, the 1985 National Acid Precipitation Assessment Plan Inventory estimates of SO^2 emissions for the mills in our sample, for example, were 200,000 tons lower than the 1984 values listed above [EPA. 1988].

Reference

U.S. Environmental Protection Agency. 1988. "Anthropogenic Emissions Data for the 1985 NAPAP Inventory." EPA-600/7-88-002. Research Triangle Park, NC.

Appendix 4-B

PULP-PRODUCING MILLS EVALUATED IN THIS STUDY

1973 and 1984

The first step in the analysis was to establish an inventory of mills producing pulp. The only inventory with information on air pollution emissions and control equipment for the time period of interest is EPA's National Emissions Data System. In this file a "point source" is defined for state reporting purposes as a facility at which the annual emissions of any one criteria pollutant equal or exceed 100 tons per year. Data for individual sources emitting more than 25 tons per year are included in the file for these facilities. In general, emissions from individual stacks and processes are defined within such a facility.

To select the data on pulp and paper facilities from the National Emissions Data System for analysis, we tried a number of different combinations of the Standard Industrial Classification codes and the National Emissions Data System's Source Classification Codes. Finally, we decided to select all plants with Source Classification Code designations for sulfate and sulfite pulping. These processes are the major air polluters. The National Emissions Data System does not include plants with less than 100 tons per year of air emissions, so some small sulfite mills are not in our study.

Groundwood mills are also of interest for this study, but there are no National Emissions Data System or Standard Industrial Classification codes specific to groundwood mills. We turned to the *1985 Post's Pulp and Paper Directory* [Brandes. 1984] to identify groundwood mill locations. We found 46 mills with groundwood capacity. Eighteen mills were already included in the inventory with either kraft or sulfite processes. Ten mills could not be matched with the National Emissions Data System plants. This left 18 groundwood mills to be analyzed for air pollution equipment costs in 1984. A similar matching left 15 groundwood mills to be analyzed for 1973.

EPA completed the first National Emissions Data System file in 1973, so it contains obvious errors and anomalies. We used secondary data sources to correct the most obvious errors. Secondary data sources included the *Post's 1974 Pulp and Paper Directory* [Medynski. 1973], an EPA report listing TRS emitters in the mid-1970s [EPA. 1979], and the plant inventory used in the water pollution analysis. As a result, the 1973 snapshot contains 150 pulp mills. Some of the mills, particularly sulfite mills, are combination mills having more than one pulping process, such as sulfite and groundwood.

The 1984 National Emissions Data System file appears to be more complete than the 1973 file, both in coverage of mills in the industry as well as records about emission sources within plants. The above-mentioned secondary data sources were used again to correct obvious errors. The resulting 1984 snapshot contains 157 pulp mills. Appendix B presents a list of the pulp mills included in our analyses.

The second step was to identify major sources of emissions at each mill. A review of the unit level National Emissions Data System records revealed incomplete data for many mills. To correct the situation to some degree, we assumed that each mill had at least one of the standard components for an identified pulping process (Table 4-B-1). For instance, each kraft mill must have some combination of digesters, evaporators, recovery furnaces, smelt dissolving tanks, lime kilns, and power boilers. There can be several such components, but there must be at least one of each in order to produce the indicated output of pulp.

The third step was to identify the daily operating rate or maximum design rate for each pulping process, measured in air-dried tons of unbleached pulp. Because the data in the National Emissions Data System files often did not balance for the different processes in a mill, we relied on the 1974 and 1985 *Post's Pulp and Paper Directories* to recompute the maximum rates in the National Emissions Data System. This step insured that individual process capacities would sum to the total required for the pulp output at each mill. To compute annual production, we assumed that all mills operated 345 days per year.

The fourth step was to identify the pollution control equipment in place for each process and the level of emission control. The various sources within a pulp and paper mill can be controlled with a wide variety of equipment. Appendix 4-C contains specific information about the technologies commonly available. Although they represent a broad spectrum of control technology, other technologies are possible. This study assumes that technologies not shown in Appendix 4-C have levels of emission control (and costs) similar to those of the most common technology applied to that source.

Although the National Emissions Data System file served as the basis for determining control technology and level of emission control, we developed a series of corrections and defaults for unavailable data. Table 4-B-1 lists the default values for control type and efficiency for each source of TSP emissions as well as for other air pollutants included in this study.

The fifth step was to determine whether process units in existing plants were in compliance with their emission limits. For this study we chose EPA's Compliance Data System as an indicator of whether plants were in or out of compliance. A request of the compliance status for pulp and paper mills was made for December 1984. The Compliance Data System Quick Look Report [Smith. 1986] lists plants and their compliance status. Plants listed as being in violation by the system were selected as candidates for further review. This preliminary screen identified 38 plants with potential compliance problems for TSP.

Because no information was provided in the Compliance Data System about why each plant was in violation, we made inquiries to compliance personnel at state air pollution control agencies to obtain this information. Through these inquiries, we found that not all the mills listed in the Compliance Data System as being in violation were truly in violation. Twenty-one of the plants were either in compliance or were noncomplying for administrative reasons (i.e., inadequate reporting or malfunctioning equipment). State personnel could not provide compliance information for 10 plants. We prepared cost estimates and emission reductions for the remaining seven plants, using the information provided by state agency personnel on what action was needed to reach compliance.

1994

Establishing the background data for snapshot year 1994 followed a sequence slightly different from the procedure used for 1973 and 1984.

The first step for the 1994 analysis was an estimate of total industry capacity in 1994. This 1994 estimate combined projections from the American Paper Institute for the near years (1985-1990) and from paper products demand based on GNP for the out years (1990-1994) [API. 1987]. The 1994 capacity was allocated to expansion at existing mills (60 percent of the new capacity) and at 20 new mills (40 percent of the new capacity). Only the six using kraft pulping were included in the analysis. Capacity was decreased at some mills during this period. We included 163 mills in the 1994 analysis.

Table 4-B-1. Required Plant Components and Control Equipment and Air Pollutant Default Values

Pulp Process	Components	Air Pollution Control Equipment	Pollutant	Control Efficiency Existing	New
Kraft (Sulfate)	Digester	Incineration	TRS	N/A	N/A
	Evaporator	Incineration	TRS	N/A	N/A
	Recovery Furnace	Scrubber ESP	TSP	95	99
		Black Liquor Oxidation	TRS	N/A	N/A
	Smelt Dissolving Tank	Demister Scrubber	TSP	80	95
	Lime Kiln	ESP Scrubber	TSP	90	99
			TRS	N/A	N/A
	Power Boiler	Cyclone ESP Scrubber	TSP	85	99
		--	SO$_2$	--	--
Sulfite	Blow Pit	--	SO$_2$	--	--
	Acid Plant	--	SO$_2$	--	--
	Power Boiler	Cyclone ESP	TSP	85	99
		--	SO$_2$	--	--
Ground-wood	Power Boiler	Cyclone ESP	TSP	85	99
		--	SO$_2$	--	--

N/A = Not applicable.

The next step for the 1994 analysis was to establish the number of emission sources at expanded or new mills. Even though existing mills could expand production somewhat without building any new process units, the assumption made for estimating how plants would be configured in 1994 was that new kraft or sulfite process equipment and associated power boilers would be built in the appropriate size to meet any increased production demands. Thus, if a plant's kraft pulping capacity was expected to increase by 500 tons per day from 1984 to 1994, then we assumed a digester, an evaporator, a recovery furnace, a smelt dissolving tank, and a lime kiln, all with capacities consistent with 500 tons per day of air-dried unbleached kraft pulp, would be built at the plant. Likewise, we assumed power boilers would be constructed to meet average industry energy requirements of 22 million BTUs per ton of paper, less energy generated in recovery boilers. Coal and wood/bark are expected to be the predominant new boiler fuels in the pulp and paper industry, so we estimated new boiler capacity at each plant to be 70 percent coal fired and 30 percent wood/bark fired.

References

American Paper Institute (API). 1987. "Paper Paperboard Wood Pulp Capacity." New York, NY.

Brandes, Debra A. (ed.). 1984. *1985 Post's Pulp and Paper Directory*. Miller Freeman Publications, San Francisco, CA.

Medynski, Ann L. (ed.). 1973. *Post's 1974 Pulp and Paper Directory*. Miller Freeman Publications, San Francisco, CA.

Smith, Franklin. 1986. "CDS Quick Look Report." Computer printout from the U.S. Environmental Protection Agency, Washington, DC.

U.S. Environmental Protection Agency. 1979. "Kraft Pulping Control of TRS Emissions from Existing Mills." EPA-450/2-78-003b. Research Triangle Park, NC.

Appendix 4-C

AIR POLLUTION CONTROL COST ESTIMATION (TSP)

We used process-specific cost functions for each of the components identified in Table 4-B-1. For each of the components, we characterized existing and new source emission reductions for TSP.

Data Sources

We took cost information from studies of the pulping industry conducted by Hendrickson et al. [1970], Arthur D. Little [1977], the Industrial Gas Cleaning Institute [1973, 1974], EPA New Source Performance Standards studies [1976, 1982, 1983], and EPA's industrial boiler studies [Radian. 1984].

Data on costs and values of recovered materials were prepared at different times. Capital cost data were adjusted to 1984 dollars using the Chemical Engineering Plant Cost Index [1985]. Values of recovered materials (soda ash and lime) were taken from the *Pulp and Paper Magazine* [Schockett. 1984]. Annual operating costs were adjusted from varying time periods by means of the Engineering News Record Construction Cost Index [1974, 1985].

Cost Functions

The process-specific cost functions derived from the above data sources are listed in Tables 4-C-1, 4-C-2, and 4-C-3.

Four types of costs -- capital, operation and maintenance, recovery credits, and annual -- were computed in each year for each component in-place as appropriate. The first two types of costs, capital and operation and maintenance, were computed using engineering cost functions for each component in the standard form:

$$\text{Cost} = a * X^b$$

where: $X =$ the average daily pulp production;
             $a =$ a linear scaling factor;
             $b =$ a scaling factor for economies of scale in
                    pollution control technology.

The recovery credits were calculated based on the amount of chemical recovery. Because recovery credits are actually savings in costs, they must be subtracted from operation and maintenance costs to give a true picture of annual costs. The annual costs were computed by adding the annual operating costs to the annualized capital costs, based on amortizing the capital costs over 15 years at a 10 percent discount rate, and subtracting the value of recovered materials.

References

Arthur D. Little. 1977. "Economic Impacts of Pulp and Paper Industry Compliance with Environmental Regulations." EPA-230/3-76-014-1. Cambridge, MA.

Chemical Engineering. 1985. "Economic Indicators." *Chemical Engineering*, December 9 and 23.

Engineering News Record. 1974. "ENR Record of Annual Forecasts 1966-1974." *Engineering News Record*, December 19, 1974.

_____. 1985. "ENR Market Trends." *Engineering News Record*, September 5, 1984.

Hendrickson E. R., J.E. Robertson, and J.B. Koogler. 1970. "Control of Atmospheric Emissions in the Wood Pulping Industry." Volume 1. Report to the Public Health Service by Environmental Engineering, Inc., Gainesville, FL.

Industrial Gas Cleaning Institute. 1973. "Air Pollution Control Technology and Costs in Seven Selected Areas." EPA-450/3-73-010. Stamford, CT.

_____. 1974. "Air Pollution Control Technology and Costs: Seven Selected Emission Sources -- Kraft Mill Recovery Boilers, Ferroalloy Furnaces, Feed and Grain Processing, Glass Melting Furnaces, Crushed

Stone and Aggregate Industry, Asphalt Saturators, and Industrial Surface Coatings." EPA-450/3-74-006. Stamford, CT.

Radian Corporation. 1984. "Industrial Boiler SO_2 Cost Report." EPA-450/3-85-011. Research Triangle Park, NC.

Schockett, Barry. 1984. "Pulp/Chemical Use, Prices Continue to Strengthen in 1984-1985." *Pulp and Paper Magazine*, 58:11:77-85.

U.S. Environmental Protection Agency. 1976. "Standards Support and Environmental Impact Statement: Proposed Standards of Performance for Kraft Pulp Mills." EPA-450/2-76-014a. Research Triangle Park, NC.

_____. 1982. "Control Techniques for Particulate Emissions from Stationary Sources." EPA-450/3-81-005b. Research Triangle Park, NC.

_____. 1983. "Review of New Source Performance Standards for Kraft Pulp Mills." EPA-450/3-83-017. Research Triangle Park, NC.

Table 4-C-1. 1984 Kraft Process Particulate Control Cost Factors ($1984)

Production Process	Particulate Control Equipment	Efficiency	(a) Coefficient	(b) Exponent
		--- Capital Costs[a] ---		
Recovery Furnace	ESP[b]	>99%	15.57	0.82
(direct)	ESP	95%<CE<99%	26.85	0.709
	Scrubber	95%	6.42	0.834
	ESP	<95%	20.574	0.666
	Scrubber	90%	2.95	0.8303
Recovery Furnace	ESP	>99%	30.17	0.789
(indirect)	ESP	95%<CE<99%	28.93	0.773
	Scrubber	90%<CE<95%	13.49	0.7145
	ESP	95%	26.46	0.650
Smelt Dissolving	Demister	85%<CE<90%	4.66	0.3874
Tank	Scrubber	>90%	3.062	0.631
Lime Kiln	ESP	>98%	19.988	0.53
	Scrubber	>98%	7.582	0.534
	Scrubber	<98%	6.315	0.532
		--- Operating and Maintenance Costs[a] ---		
Recovery Furnace	ESP	>99%	1.935	0.7510
(direct)	ESP	95-99%	3.996	0.6117
	Scrubber	>95%	1.030	1.0088
	ESP	>95%	0.023	1.1796
	Scrubber	90%	1.681	0.6657
Recovery Furnace	ESP	>99%	1.713	0.7676
(indirect)	ESP	95-99%	1.555	0.7594
	Scrubber	>95%	1.017	1.0106
	ESP	95%	0.030	1.180
Smelt Dissolving	Demister	85%<CE<90%	0.919	0.093
Tank	Scrubber	>90%	0.181	0.7329
Lime Kiln	ESP	>98%	0.661	0.9433
	Scrubber	>98%	0.153	1.0312
	Scrubber	<98%	0.117	1.0053

a. Cost equations for operating and maintenance are in the form $y = ax \exp(b)$, where x is the unit capacity in air-dried tons of pulp per day.
b. ESP = Electrostatic Precipitator.

Table 4-C-2. 1984 Sulfite Pulping Particulate Control Cost Factors ($1984)

Process	Cost Component	(a) Coefficient	(b) Exponent
Acid Plant	Capital	0.4456	0.4988
	O&M[a]	0.0140	0.8169

a. Cost equations for operating and maintenance are in the form y = ax exp(b), where x is the unit capacity in air-dried tons of pulp per day.

Table 4-C-3. 1984 Power Boiler Particulate Control Cost Factors ($1984)

Primary Fuel	Particulate Control Equipment	Efficiency	(a) Coefficient	(b) Exponent
--- Capital Costs[a] ---				
Bark or Coal	Cyclone	85%<CE<90%	12.999	0.6592
	Cyclone	90%<CE<95%	50.627	0.54
	ESP	94%<CE<99%	188.6	0.5467
	Scrubber	<90%	b	b
Oil	ESP[c]	All	d	d
--- Operating and Maintenance Costs[a] ---				
Bark or Coal	Cyclone	85%<CE<90%	0.0411	1.210
	Cyclone	90%<CE<94%	0.01158	1.597
	ESP	94%<CE<99%	4.483	0.588
	Scrubber	CE<90%	e	e
Oil	ESP	All	16.22	0.38

a. Cost equations for operating and maintenance are in the form y = ax exp(b), where x is the unit capacity in air-dried tons of pulp per day.
b. Cost equation applied is y = 80.551 + 74.850x, where x is the flue gas flow rate divided by 1000.
c. ESP = Electrostatic Precipitator.
d. Cost equation applied is y = 6.3x - (5.81 * 10 exp((-6))x exp (2) + 37.716, where x is the flue gas flow rate divided by 1000.
e. Cost equation applied is y = 134,700 + 150x + 435x exp(0.94), where x is the flue gas flow rate.

Appendix 4-D

AIR POLLUTION CONTROL COST COMPARISONS

We compared our real resource cost estimates with other real resource and financial cost estimates. The other real resource cost estimates are in the EPA "Cost of Clean Air and Water Report to Congress 1984" [EPA. 1984]. The two estimates of financial costs are the Census' "Pollution Abatement Costs and Expenditures, 1984" [Bureau of the Census. 1986] and the paper industry's "A Survey of Pulp and Paper Industry Environmental Protection Expenditures -- 1984" [National Council of the Paper Industry for Air and Stream Improvement, Inc. 1985]. We did not include the paper industry survey in our comparison because it did not report operation and maintenance or annualized capital costs; it reported only capital investments.

The "Cost of Clean Air and Water Report to Congress 1984" estimated the annualized capital, operation and maintenance, and annual costs for the pulp and paper industry. These costs were obtained by updating for inflation a 1977 EPA consultant report, "Economic Impacts of Pulp and Paper Industry Compliance with Environmental Regulations" [Arthur D. Little. 1977]. These cost estimates were updated to 1984 dollars and are presented in Table 4-D-1 as the EPA "Cost of Clean" report. The annual costs for air pollution control from the Cost of Clean model are only 70 percent of the EPA cost estimates for annual costs. The Cost of Clean model costs are lower because we used a lower capital recovery factor to annualize to capital costs and assigned higher credits for recovered materials, which lowered the net operation and maintenance costs.

The Cost of Clean model real resource cost estimate is higher than the financial cost of pollution control to the industry according to the Census data. Financial costs reflect how tax provisions, such as accelerated depreciation, reduce the actual costs of pollution control to industry. The tax provisions account for the difference between the annual capital cost estimate in the Cost of Clean model and the annual capital cost estimate as reported by Census. The annual operation and maintenance costs from the Cost of Clean model are lower than those in the Census survey because we assigned higher credits for recovered materials.

Table 4-D-1. Aggregate Cost Comparisons ($1984 10^6)

Cost Category	EPA	Model	Census[a]
Capital In-Place	$1,200	$1,270	-
Investment	$80	-	$130
Annual Capital	$240	$170	$70
Annual O&M	$170	$120	$190
Annual Costs	$410	$290	$260

a. Census investment and annual cost figures were taken from Table 4a in the Census publication. Annual capital and annual O&M figures were interpolated from Table 5a, and recovery credits were taken from Table 5a.

References

Arthur D. Little. 1977. "Economic Impacts of Pulp and Paper Industry Compliance with Environmental Regulations." EPA-230/3-76-014-1. Cambridge, MA.

National Council of the Paper Industry for Air and Stream Improvement, Inc. (NACASI). 1985. "A Survey of Pulp and Paper Industry Environmental Expenditures -- 1984." Special Report No. 85-04. New York, NY.

U.S. Bureau of the Census. 1986. "Pollution Abatement Costs and Expenditures." MA-200 (84)-1. U.S. Government Printing Office, Washington, DC.

U.S. Environmental Protection Agency (EPA). 1984. "Cost of Clean Air and Water Report to Congress 1984." EPA-230/5-84-008. Washington, DC.

Appendix 4-E

LOADINGS DATA

Table 4-E-1. Total TSP Emissions for 60 Pulp-Producing Mills

Mill	TSP Emissions (tons) 1973	1984 w/o CAA	1984 w/ CAA	Mill	TSP Emissions (tons) 1973	1984 w/o CAA	1984 w/ CAA
700	1,250	2,750	1,950	761	2,560	5,130	2,560
702	7,200	7,560	6,380	764	3,480	3,720	1,970
703	980	1,440	440	766	1,500	1,790	760
704	3,060	6,780	1,070	770	3,230	4,360	400
705	2,860	3,180	2,110	772	7,620	7,940	850
708	5,130	6,410	720	776	3,200	2,800	610
712	1,490	1,680	880	786	220	240	50
713	2,940	3,300	1,430	788	3,010	2,020	80
716	3,570	6,230	580	791	3,530	3,680	1,290
718	910	910	340	795	3,730	4,240	1,160
720	10,880	7,500	880	798	2,760	2,720	2,220
723	3,310	2,990	1,530	804	5,670	5,670	160
724	1,860	2,790	580	808	1,110	1,290	1,110
725	2,650	9,270	2,020	814	1,800	2,180	1,620
729	1,830	2,130	920	817	2,720	1,670	680
730	1,020	1,040	860	820	2,260	2,350	660
731	1,110	790	50	822	380	1,240	840
735	4,980	6,500	940	823	680	830	240
736	6,760	8,840	3,980	827	300	300	10
737	2,580	2,920	1,570	828	430	530	280
739	1,440	4,110	650	835	7,020	6,870	1,780
741	3,150	3,320	950	837	2,420	1,570	100
742	1,520	1,830	1,350	841	9,600	16,000	8,830
743	12,650	12,650	5,040	855	3,650	3,960	580
744	2,730	3,640	150	858	1,970	1,210	580
747	3,050	3,410	850	860	2,470	2,890	1,240
748	2,300	2,300	250	864	920	990	390
751	3,540	4,430	1,750	868	2,840	3,830	240
754	3,440	4,210	510	880	790	1,620	150
759	440	600	320				

Appendix 4-F

AIR QUALITY MODEL

The Model City Program (MCP) was developed for EPA to provide a fast and affordable method to estimate the ground-level impact for large numbers of sources and for many areas across the United States [Maxwell and Manuel. 1985]. The objective of developing the MCP model was to mimic the performance of the Industrial Source Complex air quality dispersion model using a microcomputer. To model air quality changes with the MCP model, its developers made the following approximations:

- A fixed set of representative source characteristics (i.e., stack height, stack diameter, effluent temperature, exit velocity, and source dimensions) are used to approximate real sources.

- Three particle-settling velocities are used to represent the full range of particle sizes emitted by a source and their settling velocities.

- A data-reduction methodology is used that sets equal to zero all concentrations less than 10 percent of the maximum concentration for each representative source and each time period modeled.

- All sources are located at the center of grid squares.

The MCP model is essentially a database management system. The MCP database used in this analysis simulates ground level concentrations using meteorological data for 55 cities. For each city, the MCP database contains the annual average spatial patterns of pollutant for each source in a set of 32 representative sources. The set of representative sources is the same for each city and includes point and area sources. The spatial patterns were determined using the Industrial Source Complex Long-Term (ISCLT) air quality dispersion model. The ISCLT inputs included multi-year stability array (STAR) data for each city developed by Martin and

Tikvart [1968], appropriate average temperature data for each city, and average mixing heights determined from Holzworth's [1972] monograph.

To use the MCP model, an analyst enters data describing the real sources into the model. For example, for a point source, the analyst enters the source location, stack height, stack diameter, effluent temperature, exit velocity, and emission rate. The model uses this information to calculate the effective stack height for the real source using preselected meteorological conditions. (The effective stack height is the source stack height plus the plume rise that results from the momentum and buoyancy of the gas being exhausted through the stack.) To represent the real source, the MCP model selects the representative source with the effective stack height of the real source. This process is repeated for each real source.

Complex sources with multiple emission points are handled by examining each emission point separately. The model will select the representative source that best represents each emission point. Sources that emit varying sizes of particles are handled similarly. The MCP set of representative sources includes sources that are identical except for the settling velocity used to represent the size of the particles emitted. (Settling velocities are proportional to the particle density.) Thus, the model allows for segregating particles into three categories: gases and small particles (particles with diameters less than 12 μm), medium particles (particles with diameters between 12 and 24 μm), and large particles (particles with diameters greater than 24 μm).

Once the representative sources have been selected, the model positions each source on a base grid. For each source, the model retrieves from the data base the ground level concentrations for the source. It then adjusts them to reflect the user-specified emission rate for the real source to estimate the aggregated ground level concentrations at each of the 1,600 MCP receptors. The 1,600 receptors are arranged in a grid of 40 rows by 40 columns spaced 1-km apart.

Application of the MCP Model to Pulp and Paper Mills

We used the MCP model to estimate ground level concentrations of TSP due to the TSP emitted by 60 pulp and paper mills across the United States. The following assumptions were used in the application of the model:

1. For each plant, all sources were collocated.

2. The assumptions that are applicable to the Industrial Source Complex model are applicable to the MCP model (i.e., flat terrain, emissions constant throughout the year).

3. All TSP emissions were assumed to be small particles.

4. The emission rate for each source in grams per second was calculated by distributing the annual total TSP emitted (in tons) equally throughout the year.

5. Only sources included in the National Emissions Data System inventory were modeled [EPA. 1984].

6. Sources with stack heights, stack diameter, temperature (°F), and exit velocity equal to zero in the National Emissions Data System data base were modeled as volume sources with dimensions of 250m x 250m (the MCP default dimensions).

7. The modeled source characteristics (emission rate, stack height, stack diameter, effluent temperature, and exit velocity) were taken from the 1984 National Emissions Data System inventory. These parameters were for the with and without Clean Air Act scenarios.

Historically, the pulp and paper industry has strived to recover a large percentage of the product that would be emitted to the atmosphere. Consequently, before the Clean Air Act, the "common" pulp and paper mill used emissions control devices. In addition, with the exception of the lime kilns, about 85 percent of the TSP emissions from processes within a pulp and paper mill before controls are small (<10 μm) particles [EPA. 1985]. Adding particulate emissions control devices typically results in the proportion of small particles being emitted to increase because most of the particulate emissions control devices are more efficient at removing large particles than small particles. The typical control device added to the lime kilns results in the percentage of small particles emitted from these sources to represent about 90 percent of the mass emitted. Because the controlled emissions from pulp and paper mills are mostly small particles, all TSP emissions were assumed to be small particles (<12 μm).

References

Holzworth, George C. 1972. "Mixing Heights, Wind Speeds, and Potential for Urban Air Pollution Throughout the Contiguous United States." AP-101. U.S. Environmental Protection Agency, Research Triangle Park, NC.

Martin, Dennis O., and Joseph A. Tikvart. 1968. "A General Atmospheric Model for Estimating the Effects on Air Quality of One or More Sources." Paper presented at the 61st Annual Air Pollution Control Association Meeting, St. Paul, MN.

Maxwell, Christopher D., and Ernest H. Manuel Jr. 1985. "Description of the Model City Program Methodology and Development." Report to the U.S. Environmental Protection Agency. Versar, Inc., Columbia, MD.

U.S. Environmental Protection Agency. 1984. "National Emissions Report." Office of Air Quality Planning and Standards, Research Triangle Park, NC.

_____. 1985. "Compilation of Air Pollutant Emission Factors: Stationary Point and Area Sources." AP-42. Office of Air Quality Planning and Standards, Research Triangle Park, NC.

Appendix 4-G

EMISSIONS VERSUS CONCENTRATIONS

The mills with the largest emission reductions were not necessarily the mills with the largest reductions in ground-level TSP concentrations. Table 4-G-1 lists, in order, the 20 mills with the largest emission reductions and, in order, the 20 mills with the largest reductions in TSP concentrations.

The mills with very large reductions in TSP emissions usually are the very large mills that have stack parameters (stack height, stack diameter, effluent temperature, and exit velocity) that result in very high effective stack heights. (The effective stack height is the source stack height plus the plume rise that results from the momentum and buoyancy of the gas being exhausted through the stack.) Consequently, the plumes from these mills are highly diluted before they reach the ground. Mills with moderate reductions in emissions and stack parameters that result in low effective stack heights may show large changes in ground-level TSP concentrations because their plumes are not diluted as much as the mills with larger effective stack heights. This result indicates the importance of considering source-receptor relationships in estimating air quality impacts, which could not be done with linear rollback methods.

Table 4-G-1. Mills with Large Reductions in TSP Emissions and
Mills with Large Reductions in TSP Concentrations

Large Reductions in Emissions		Large Reductions in Concentrations	
Mill	Annual Reduction in Emissions (tons)	Mill	Maximum Reduction in Concentrations[a] ($\mu g/m^3$)
743	7,610	863	122
725	7,260	731	87
841	7,170	776	67
722	7,100	855	58
720	6,620	736	51
704	5,720	857	40
708	5,700	856	33
716	5,640	868	26
735	5,560	791	15
819	5,320	864	14
835	5,100	724	13
736	4,870	708	12
770	3,960	770	12
754	3,700	772	11
868	3,590	748	11
821	3,500	848	10
863	3,500	754	9
737	3,460	720	7
855	3,380	788	7
734	3,230	739	7

a. The difference between the maximum annual average TSP concentration predicted by the MCP model assuming emissions in 1984 without the Clean Air Act and the maximum concentration predicted assuming emissions in 1984 with the Clean Air Act.

Appendix 4-H

MODEL VALIDATION

EPA [1986] provides a good discussion of the accuracy and uncertainty of air quality dispersion models. Long-term average concentration estimates (e.g., annual average concentrations) are more reliable than short-term concentration estimates, and studies have shown that estimates of the maximum short-term concentrations typically have errors of plus or minus 10 to 40 percent -- well within the factor-of-two accuracy that has long been recognized for these models.

The majority of reports evaluating the performance of the Industrial Source Complex model have focused on the short-term version of the model. One study that evaluated the performance of the long-term model used it to estimate measured values [Heron et al. 1984]. The estimated and measured annual average TSP concentrations were compared using regression analysis. The resulting regression equation was:

$$\Delta TSP_o = 1.18 \, \Delta TSP_p + 20.2$$

where TSP_o is the observed TSP concentration and TSP_p is the ISCLT predicted TSP concentration.

The intercept, 20.2 $\mu g/m^3$, is an estimate of the background TSP concentration due to natural and other sources not included in the modeling. This estimate for the background concentration was in good agreement with the value of 25 $\mu g/m^3$ estimated by Minnesota Pollution Control Agency for the facility modeled. After correcting the Industrial Source Complex modeled estimates using the regression equation, the correlation coefficient of the observed and predicted TSP concentrations was 0.97 and was significant at the one percent level.

The primary objective of the MCP model is to mimic the performance of the Industrial Source Complex model. As part of the development of the MCP model, the MCP and the Industrial Source Complex models were used to estimate TSP concentrations in and around the Baltimore, Maryland, area [Maxwell and Mahn. 1988]. In addition, the MCP estimates were compared with measured TSP concentrations. The results of these comparisons for annual average concentrations were:

1. At all except four of the 1,600 receptors, the MCP estimates were within five $\mu g/m^3$ of the Industrial Source Complex predicted values. At the vast majority of receptors, the difference was less than three $\mu g/m^3$, and only one receptor differed by more than 10 $\mu g/m^3$.

2. The MCP estimates exhibited a downward bias of about one $\mu g/m^3$ as a result of a data reduction methodology that ignores contributions of any source that provides less than 10 percent of the maximum concentration. The bias causes the MCP estimates to be slightly lower than the Industrial Source Complex estimates.

3. The MCP model replicated the measured TSP concentrations in suburban and rural areas. However, in urban and industrial areas, the MCP model underestimated the measured values. This bias was attributed to the lack of a good estimate of the background concentrations for the urban and industrial areas.

As part of this study, air quality estimates for 1984 emissions from 14 pulp and paper mills were compared with previously completed Industrial Source Complex modeling estimates for the same 14 mills [Maxwell et al. 1987]. It was not possible to compare the MCP model and Industrial Source Complex model estimates on a point-by-point basis because the MCP model uses a rectangular grid of receptors, whereas the Industrial Source Complex model uses a polar receptor grid arrangement. Consequently, the two models did not predict for exactly the same locations in space. In addition, it is probable that different years of meteorology were used in the MCP and Industrial Source Complex models because the actual years and the location of the meteorological stations used for Industrial Source Complex modeling were not recorded. (For the MCP modeling, statistical summaries of five years of meteorological data were nominally used.)

Table 4-H-1 lists the maximum annual average TSP concentration predicted by the MCP and Industrial Source Complex models for the 14 mills. The Industrial Source Complex modeling had receptors located at distances as close as 200 meters and 500 meters from the plants. However, because the MCP model predicts concentrations no closer than 707 meters from the sources, only Industrial Source Complex receptors located at least one km were used in the comparison. In general, the Industrial Source Complex maximum concentrations are larger than the MCP maximum concentrations. This results primarily from the greater density of Industrial Source Complex receptors close to the plant site as compared to the MCP receptor grid. Overall, the MCP estimates, however, are in reasonable agreement with the Industrial Source Complex results given the difference in receptor locations and the probable difference in meteorology used by the two models.

Table 4-H-1. Maximum Annual Average TSP Concentrations ($\mu g/m^3$)

Mill	State	ISCLT	MCP	ISCLT[a]
703	AL	0.91	0.86	
704	AL	1.62	1.12	
708	AL	2.09	1.60	
722	AR	0.30	0.29	
723	AR	1.21	1.17	
731	CA	0.36	0.16	
748	GA	1.32	1.25	1.17
744	GA	0.32	0.16	0.16
759	LA	3.23	1.43	1.13
770	LA	1.08	1.18	
827	PA	0.15	0.10	
837	SC	0.04	0.04	
852	VA	5.52	2.90	2.67
879	WI	0.37	0.41	

a. ISCLT re-estimate using the same source data, meteorological data, and receptor grid used in the MCP modeling.

Four of the mills in Table 4-H-1 were selected for re-estimating the air quality about the mills by reapplying the Industrial Source Complex Long Term model using the source data, meteorological data, and receptor set used in the MCP modeling. Thus, the only difference between the Industrial Source Complex Long Term and MCP models results for these four mills are that the MCP results are based on the MCP generic sources being used in lieu of the real source characteristics used in the Industrial Source Complex Long Term modeling. The re-estimated maximum annual average TSP concentrations for these four mills are also shown in Table 4-H-1. For three of the four mills, the re-estimated Industrial Source

Complex Long Term model estimated maximum annual average TSP concentrations are more similar to the MCP model estimates than were the previous Industrial Source Complex Long Term modeling results. The one mill for which the re-estimated value is less similar that the original estimate was for a mill for which the original estimate was already very similar to the MCP model estimate.

References

Heron, Thomas M., James F. Kelly, and Paul G. Haataja. 1984. "Validation of the Industrial Source Complex Dispersion Model in a Rural Setting." *Journal of the Air Pollution Control Association*, 34:365-369.

Maxwell, Christopher D., Ducan McHale, Sally A. Mahn, and Keith Timbre. 1987. "Sensitivity Analysis of the Model City Program." Report to the U.S. Environmental Protection Agency. Versar, Inc., Columbia, MD.

Maxwell, Christopher D., and Sally A. Mahn. 1988. "Assessment of the Performance of the Model City Program." Report to the U.S. Environmental Protection Agency. Versar, Inc., Columbia, MD.

U.S. Environmental Protection Agency. 1986. "Guideline on Air Quality Models (Revised)." EPA-450/2-78-027R. Research Triangle Park, NC.

Appendix 4-I

DERIVATION OF CONCENTRATION RESPONSE FUNCTIONS FOR HEALTH EFFECTS

Mortality

In many of the previous EPA benefit analyses for TSP, the mortality effects were estimated using concentration-response functions developed by Mazumdar et al. [1980]. The Mazumdar study examines the relationship between daily variations in mortality rates and changes in British Smoke and SO_2 concentrations. Data for the study were available for 14 winters in the London area.

More recently, Schwartz and Marcus [1990] have reanalyzed the London data using somewhat different statistical techniques. This study is the basis of the reduced mortality benefits calculated here. Their effort reports results for several different specifications. In this report, changes in mortality are evaluated for a specification in which British Smoke values are restricted to measurements less than 200 $\mu g/m^3$ and the effects of both temperature and humidity are controlled for. In differential terms, the basic relationship is:

$$\Delta \text{ LMORT} = 0.138 \text{ x } \Delta \text{ BS}$$

where: Δ LMORT = change in daily mortality rates in London
Δ BS = change in daily concentrations of British Smoke

There are three adjustments to be made to this relationship before it can be applied. First, it is necessary to convert British Smoke measures to TSP. This is accomplished using the result that a 1 $\mu g/m^3$ change in British Smoke is equivalent to a 1.41 $\mu g/m^3$ change in TSP when British Smoke is less than 200 $\mu g/m^3$. Second, the equation must be standardized to a unit population value. This can be done by dividing the mortality estimate by eight million, the approximate population of London during the period of analysis. Finally, the equation is adjusted from a daily mortality rate equation. However, it is probably not appropriate to develop an annual counterpart. Because the statistical equations included data from the winter months only, we used mortality rates from the heating season only (November through February).

With these changes, the coefficient of the Schwartz and Marcus equation, B, is 1.464×10^{-6}, with a range of 1.068×10^{-6} to 1.932×10^{-6}. This range represents two standard deviations below and above the point estimate, respectively. Statistical lives saved are computed as:

$$\Delta \text{ MORT } = \text{ B x } \Delta \text{ TSP HEAT x POP}$$

where: Δ MORT = change in annual mortality rate
            B = the imputed coefficient ranging from 1.068×10^{-6} to 1.932×10^{-6}
Δ TSP HEAT = the change in mean TSP (during heating season)
       POP = the exposed population.

Thus, for a 1 $\mu g/m^3$ change in TSP and a population of one million, the expected number of lives saved during the heating season is one and one-half. Annual mean TSP is used as a proxy for daily mean TSP during the four month heating season, with a factor adjustment for the count of days in the heating season.

Morbidity

The best available study of the relationship between morbidity and TSP concentrations is Ostro [1987]. Ostro uses morbidity data from the Health Interview Surveys for 1976-1981. He combines these data with various demographic measures and data on fine particle concentrations developed from airport visibility data. Fine particle data are thought to be the appropriate relevant measure of pollution for TSP-related morbidity effects. A Poisson distribution is used to model several health effects including work-loss days (WLDs), reduced activity-days (RADs), and respiratory-related reduced activity days (RRADs). Ostro also used a fixed-effects specification to control for unobserved differences in the cross-sectional data.

The concentration-response functions estimated by Ostro regressed a measure of the individual's acute WLDs, RADs, or RRADs against the fine particles measure and his personal, economic, and other characteristics. The measure of fine particles was a lagged 2-week average used to represent the 2-week exposure period before the period under consideration by the Health Interview Survey. In addition to the fine particle measure, the variables included the individual's age, sex, race, education, family income, marital status, existence of a chronic condition, and average 2-week minimum temperature. The model was estimated for each of the six years, 1976 through 1981.

Using the point estimate of fine particles in Ostro's 1979 RRAD equation, the results indicate that a 1 $\mu g/m^3$ change in the 2-week average of fine particles will lead to a one percent change in RRADs. Assuming a 2-week observed RRAD equal to 0.39 (obtained from Ostro's 1979 data), the above change in fine particles will lead to a decrease in RRADs of 0.0039 days over the following 2-week period.

The evaluation of the change in RRADs requires information on fine particles and the baseline (i.e., current) level of RRADs. The measure of fine particles was obtained by multiplying TSP data by 0.26. Baseline RRADs were obtained from published sources.

Benefit estimates for the Ostro study are divided into three categories: reductions in illness occurring during work time (WLDs); reductions in illness occurring during leisure time (RADs); and, reductions in direct medical expenditures (DMEs) associated with reductions in illness. The algorithms for each of these components, expressed in terms of the ith geographic unit of observation, are summarized below. Because of the extensive evidence linking TSP exposure to respiratory illness, the benefit algorithms are limited to Ostro's RRAD concentration-response function.

The reduction in respiratory-related WLDs associated with a fine particle reduction is calculated from:

$$\Delta \text{RWLD}_i = (b_j \text{ RRAD} \times \Delta \text{FP}_i) \times 0.47 \times \text{EMP}_i \times 26$$

where: ΔRWLD_i =   change in respiratory-related work-loss days;

b_j =   coefficients of fine particles in the RRAD equations (point estimate equal to 0.01 with a range of 0.0067 to 0.0221);[1]

RRAD =   average number of respiratory-related RADs over a 2-week period;

ΔFP_i =   the change in the 2-week average of fine particles in $\mu g/m^3$;

0.47 =   percent of a worker's RADs that are also WLDs;

EMP_i =   number of employees;

26 =   factor to convert 2-week change to an annual change;

i = geographic index;

j = minimum, midpoint, or maximum estimate. index.

The reduction in RRADs net of RWLDs that results from a reduction in fine particles is equal to:

$$\Delta \text{RRAD}_i = (b_j \text{ RRAD} \times \Delta \text{FP}_i)(\text{NONWORKERS}_i + 0.53 \text{ EMP}_i) \times 26$$

where: ΔRRAD_i = change in respiratory-related restricted-activity days;

NONWORKERS$_i$ = nonemployed population;

0.53 = percent of a worker's RAD occurring during leisure time.

The other variables are as defined before.

DME benefits are calculated from:

$$\text{DME\$}_i = \frac{\Delta \text{RRAD}_i + \Delta \text{RWLD}_i}{26 \times \text{RRAD}} \times \$70.00 \times \text{POP}_i$$

where: DME$_i$ = change in direct medical expenditures associated with acute respiratory illness;

\$70.00 = per capita direct medical expenditures on acute respiratory illness in 1984 dollars;

POP$_i$ = population in i^{th} area.

Household Soiling

Household benefits of reduced soiling and damage to materials were derived from a model of consumer behavior formulated and estimated by Mathtech [1982]. Since the model estimates economic damages directly, without an intermediate calculation of physical damage, we defer discussion to the section on valuation of soiling and materials damage.

Note

1. The range represents the lowest and highest significant coefficient estimates obtained for the yearly equations estimated by Ostro.

References

Mathtech, Inc. 1982. "Benefits Analysis of Alternative Secondary National Ambient Air Quality Standards for Sulfur Dioxide and Total Suspended Particulates." Volume II. Report to the U.S. Environmental Protection Agency. Princeton, NJ.

Mazumdar, S., H. Schimmel, and I. Higgins. 1980. "Relation of Air Pollution to Mortality: An Exploration Using Daily Data for 14 London Winters, 1958-1972." Report to the Electric Power Research Institute, Palo Alto, CA.

Ostro, Bart D. 1987. "Air Pollution and Morbidity Revisited: A Specification Test." *Journal of Environmental Economics and Management*, 14:87-98.

Schwartz, Joel, and Allan Marcus. 1990. "Mortality and Air Pollution in London: A Time Series Analysis." *American Journal of Epidemiology*, 131:185-194.

Chapter 5

BENEFITS APPROACH

A few of the provisions of the Clean Air Act mandate benefits-based standards as the regulatory approach for reducing emissions. Because the Clean Air Act requires setting standards in these cases based on balancing benefits and costs, there should be a close correspondence between the benefits from improvements in human health and welfare and the costs of emission reductions. This correspondence should result in: (1) aggregate benefits from the regulation exceeding aggregate costs, and (2) benefits exceeding costs at most mills. The absence of air quality modeling and formal benefit-cost comparisons in implementing the benefits-based standard for total reduced sulfur (TRS) limits the likelihood of achieving a reasonable balance in all circumstances, however. Consequently, the relationship between benefits and costs is likely to vary across unique geographic situations, but not nearly to the extent or to the degree that it would with technology- or ambient-based standards.

An assessment of the efficiency of a benefits-based standard in reducing environmental risks requires linking the costs of implementing a specific regulatory program and the benefits that accrue from it. The linkages for regulation of TRS emissions, which are generated only by kraft pulp mills, are illustrated in Figure 5-1. This figure shows that, when implemented, a regulatory program based on a benefits-based standard will impose pollution control costs and reduce TRS emissions. These emission reductions, just as in the case of TSP emissions, will improve ambient air quality in the vicinity of kraft pulp mills. An improvement in air quality will decrease human exposure to air pollution. This will diminish adverse health and welfare effects. Benefit estimates are based on monetary valuations of people's willingness-to-pay for these improvements in health and welfare.

In a manner parallel to the previous chapter, we estimate the real-resource costs required from the kraft segment of the pulp and paper industry to comply with a benefits-based standard and the resulting reduction in TRS emissions. The costs and emission reductions result

Figure 5-1. Framework for Benefit-Cost Analysis of Clean Air Act Benefits-Based Standards

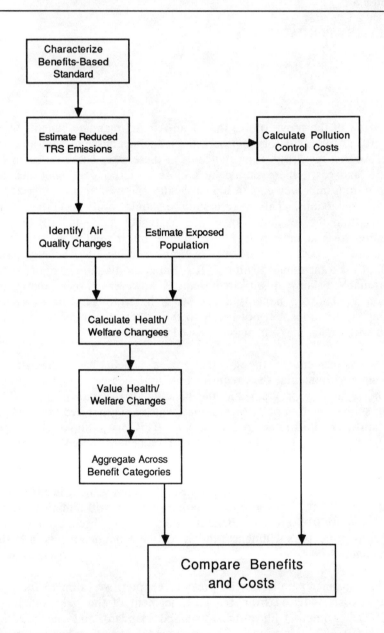

primarily from the attainment by existing kraft pulp mills of emission limits specified in state-issued operating permits.

Decreasing TRS emissions from kraft pulp mills reduces the rotten egg smell associated with pulp mills. The primary benefits of reducing TRS emissions are decreases in minor morbidity and nuisance.

POLLUTION CONTROL COSTS AND EMISSIONS

We estimate the real-resource pollution control costs and TRS emissions at three points in time - 1973, 1984, and 1994. Costs incurred before 1973 are not considered to be associated with the Clean Air Act because the regulatory requirements of the Act were not implemented until the late 1970s. Costs in 1984 are the costs for air pollution control under current regulations. Costs in 1994 are the future costs of pollution control, taking into account increased production, but assuming the same emission limits as in 1984. The other relevant type of cost and emission information is associated with 1984 production in the absence of federal regulations.

The following background section is truncated compared to the TSP background section because the information about the location and production at individual mills is repetitive of the information in Chapter 4. We assume that the reader is familiar with the approach described in that chapter. In addition, only mills that produce kraft pulp, 129 mills of the 157 mills covered in Chapter 4, are sources of TRS emissions.

Inventory -- 1973, 1984, and 1994

We identified 115 major kraft pulp mills operating in 1973 and 129 in 1984 and projected there would be 135 operating in 1994. We could find sufficient data to characterize pollution control costs and emission loadings for all these mills because of an EPA survey completed in 1976 [EPA. 1979]. We should point out, however, that only 42 mills in 1973 were incurring emission reduction costs.

We describe the procedures used for establishing this inventory in Appendix 5-A and list the mills covered in Appendix B. The background data on these mills are presented jointly for two of the three snapshot years, 1973 and 1984, and separately for 1994. The data categories for each year are emission sources within mills, daily production, pollution control equipment in place, and compliance status.

Emissions Data

EPA's National Emissions Data System does not include an estimate of TRS emissions for 1973 and 1984. We calculated TRS emissions for each mill based on TRS emission factors for each unit process, its level of control, and plant capacity [EPA. 1979].

Cost Modeling

We constructed the Cost of Clean model to estimate source-specific costs for points in time and regulatory conditions [E.H. Pechan and Associates. 1987]. Figure 5-2 illustrates the basic structure of the model with its distinct algorithms for each baseline year or condition.

Baseline Costs. The Cost of Clean model computes the capital, operation and maintenance, and annual costs of the emission reduction processes that are in place in each of the three baseline years: 1973, 1984, and 1994. Control costs for reductions in TRS emissions are different from those for TSP control in that there are no recovered materials, and all of the equipment cost is attributable to air pollution emission reductions. (See Appendix 5-B for a description of the specific cost functions.) Parallel to the computation of emission reduction costs for each of three baseline years, the model calculates the emission reductions and the resulting emission levels and reports them in both pounds of TRS per ton of pulp production and tons of TRS in the given year.

TRS control costs in 1973 are estimated by comparing plant-specific emission factors to emission limits promulgated in 1979 for existing plants. If plant emission factors are at or below the mandated levels, we estimated costs of TRS emission reductions. This assumes that mills meeting the TRS standards in 1973 are only doing so in response to stringent state or local regulations in existence at the time.

For 1984, if plant emission factors exceeded the mandated limits, we estimated costs of TRS emission reductions. Because all kraft mills would have to meet the TRS limits by 1984, costs are estimated for all sources that would have to add controls in order to be in compliance with these limits. Post-control TRS emissions are estimated using the mandated emission limits where controls are applied.

TRS control costs for 1994 combine the 1984 costs with the additional costs of expanded and new mills meeting New Source Performance Standards. Costs and emissions are estimated accordingly.

Figure 5-2. Simplified Flowchart of the Cost of Clean Model -- TRS

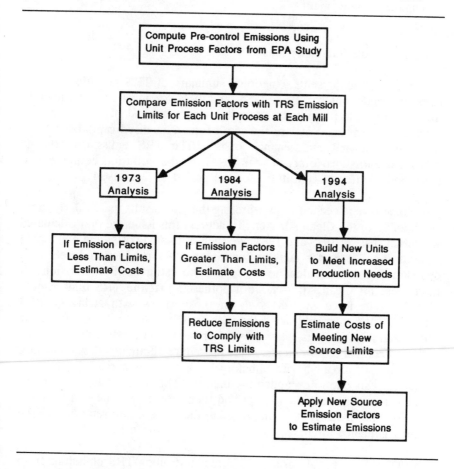

Compliance Costs. The model can assess compliance costs for emission sources not in compliance with their state operating permits.

Even though EPA's Compliance Data System and newspaper reports[1] indicated that there were compliance problems, we could not obtain sufficient information to calculate the costs of compliance for any mill.

Costs of More Stringent Regulations. The model could assess the costs to meet more stringent emission limits. We could not find documented cost information to calculate the costs of more stringent emission limits. We do estimate, however, in Chapter 7 the costs of restraining emission increases associated with expanded production based on engineering judgment.

Estimated Costs and Emissions

This section presents a national summary of TRS pollution control costs and emissions for 1973, 1984, and 1994 for most kraft pulp mills. Cost components presented are capital, operation and maintenance, and annual. The annual costs are the sum of the annualized capital costs and the operation and maintenance costs. The TRS emissions are the calculated emissions for each year based on the pollution equipment in place under different conditions for the period 1973 to 1994.

Some care is needed in interpreting the cost results, particularly those attributable to the Clean Air Act. To address this issue, we first calculated the actual costs in 1973 and 1984. We then estimated costs (and emissions) without the Clean Air Act by combining 1984 production levels with emission factors equivalent to those used to estimate 1973 emissions. If there were no Clean Air Act, we hypothesize that the TRS limits would have been in force in 1984 only in those states with stringent 1973 standards. We then subtracted the without Clean Air Act costs from the with-Clean Air Act costs, where costs represent 1984 production with 1984 pollution control requirements. This procedure eliminates those emission reduction costs already incurred before 1973, plus any increase in emission reduction costs due to expanded production. The costs attributable to the Clean Air Act in the 1973-1984 period then reflect only the real resource costs for the industry to comply with the state interpretation of the national guidance for reducing TRS emissions from existing mills.

1973. The annual costs in 1973 for controlling TRS emissions from kraft mills are listed in Table 5-1. The annual costs were approximately $40 million, slightly less than the costs associated with TSP controls. The annual costs were attributable to state and local rules. The industry emitted an estimated 60,000 tons of TRS in 1973.

Table 5-1. National Summary of Estimated TRS Pollution Control Costs
and Emissions for Kraft Mills in This Study ($1984 10^6)

Scenario	Costs			Emissions
	Capital	O&M	Annual	(tons 10^3)
1973	$195	$10	$40	60
1984 w/o CAA	$210	$15	$50	75
1984 w/ CAA	$790	$65	$165	7
1984 w/ full compliance	$790	$65	$165	7
1994	$970	$75	$205	8

1984 without the Clean Air Act. The annual costs in 1984 for
controlling TRS if the Clean Air Act requirements had not been
implemented (but with state and local requirements) would have been
approximately $50 million. The 1984 TRS emissions without the Clean
Air Act would have been about 75,000 tons (1.2 times 1973 emissions).

1984 with the Clean Air Act. The annual costs in 1984 for controlling
TRS with Clean Air Act requirements were approximately $165 million,
compared to an estimated $50 million without the Clean Air Act. This
additional investment of $115 million in TRS emission controls resulted
in a 53,000-ton decrease in TRS instead of a 15,000-ton increase in TRS
between 1973 and 1984. The average cost per ton reduced was $1,900.

1984 with Compliance. The annual costs in 1984 for controlling TRS
emissions with full compliance with the Clean Air Act could not be
estimated because of data limitations. EPA's Compliance Data System
indicates that five mills were not in compliance in 1984, but cannot
distinguish between administrative and emission exceedance violations. For
this study, we assumed that existing mills were in full compliance and,
therefore, that costs for 1984 with Compliance would be the same as costs
for 1984 with the Clean Air Act.

1994. The annual costs in 1994 to meet 1984 existing and new source
emission limits for TRS are projected to be approximately $205 million.
The 1994 annual costs would be 1.3 times the 1984 annual costs. In spite

of these additional expenditures, total TRS emissions from pulping operations would increase by 15 percent as a result of a 15 percent increase in capacity.

The costs and emission estimates do not represent the total costs of pollution control or emission reductions on the part of the entire industry. They represent most of the costs and emissions, however, based on comparisons with total industry capacity and emissions in 1973 and 1984. The mills covered in this analysis accounted for 97 percent of the 1973 kraft pulp production from the industry and 94 percent of the 1984 kraft pulp production [Medynski. 1973, Brandes. 1984].

AIR QUALITY IMPACTS

The benefits of improved air quality under the Clean Air Act must be measured by changes in specific air quality parameters. In this study, we followed EPA convention in adopting TRS as the appropriate pollutant indicator for sulfur compounds. Kraft pulping operations emit four reduced sulfur compounds -- hydrogen sulfide, methyl mercaptan, dimethyl sulfide, and dimethyl disulfide. EPA determined that an additive measure, TRS, best reflects the presence of these four compounds.

Sample Sites

As stated earlier, we modeled TRS emissions and pollution control costs for 129 mills that were producing kraft pulp in 1984. Unlike the water quality analysis, we did not need to limit the air quality analysis to those areas where kraft pulp mills were the only industrial source of TRS emissions. The air quality model allowed us to calculate the incremental change in ambient air quality concentrations without concern for other stationary sources of emissions.

We did not model the changes in ambient air quality from reductions in TRS emissions from all 129 mills, however. We eliminated 13 mills with no change in TRS emissions as a result of the Clean Air Act, 45 mills with incomplete data on both emission and cost changes, and 11 mills with less than a 10-ton reduction of TRS emissions as a result of the Clean Air Act. As a result of these constraints, we limited our TRS benefit analysis to 60 mills located fairly evenly across the four pulp- and paper-producing regions in the United States (Figure 5-3). These 60 mills account for 50 percent of the TRS emissions from kraft pulping.

Figure 5-3. Map of Kraft Pulp Mills Included in the Air (TRS) Analysis

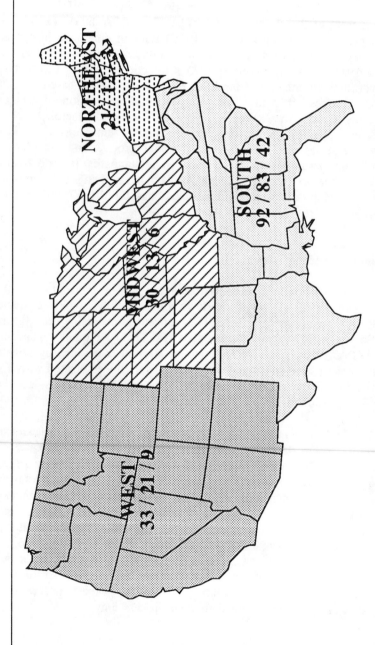

NORTHEAST
21 / 17 / 17

MIDWEST
30 / 13 / 6

SOUTH
92 / 83 / 42

WEST
33 / 21 / 9

LEGEND: No. of pulp producing mills / no. of mills with cost estimates / no. of mills in benefit-cost analysis

Scenarios

Once the mills were selected, we assembled data for estimating TRS emissions for three scenarios. Scenario 1 portrays 1973 estimated emissions from kraft pulping operations based on EPA data about process units in place at each mill and on a 1976 survey of emissions by process units. It approximates emissions discharged before the Clean Air Act by applying 1973 emission factors to 1973 production. Scenario 2 portrays 1984 hypothetical emissions from kraft pulping operations based on EPA data about process units in place. It approximates emissions discharged without the Clean Air Act by applying 1973 emission factors to 1984 production. Scenario 3 portrays 1984 estimated emissions from kraft pulping operations based on EPA data about process units in place at each mill and assuming compliance with state standards. It approximates emissions discharged with the Clean Air Act by applying 1984 emission factors to 1984 production. Emission loadings for the 60 mills for the three scenarios are presented in Appendix 5-C.

Air Quality Modeling

TRS is principally a nuisance pollutant with a pungent odor that humans can smell at relatively low concentrations. We evaluated population exposure to short-term TRS concentrations using the Multiple Point Source with TERrain Adjustment (MPTER) air quality dispersion model [EPA. 1980]. The MPTER model allowed us to identify possible combinations of wind speed and atmospheric stability associated with TRS concentrations above the odor threshold. The MPTER model is described briefly in Appendix 5-D.

The MPTER and other short-term dispersion models require extensive input data. These data requirements make it prohibitively expensive to apply the MPTER model to estimate concentrations for many sources located in many regions. Although we incorporated the actual total mill TRS emissions for each mill, we had to assume that all mills otherwise looked like a single "model" mill. The model mill consisted of eight process units: recovery furnace, smelt tank, lime kiln, digesters, multiple effect evaporators, brown stock washers, black liquor oxidation tank, and condensate stripper. Each process unit had an exhaust stack with characteristics (i.e., stack height, stack diameter, effluent temperature, and exit velocity) that remained constant across mills. A fixed percentage of mill TRS emissions was assigned to each process unit.

We selected a representative meteorological data set for each mill in the analysis. Usually, the meteorological data set comes from the National Weather Service station closest to the mill. Occasionally, it is a more distant station. For example, if the closest station were a coastal site and the mill is located inland, then we selected a more distant, but more representative inland station.

We used the MPTER model to simulate TRS concentrations at receptors located around each mill. The receptors were located along 16 wind direction radials and at each of five distances: 0.5, 1.5, 2.5, 4.0, and 7.5 miles. The distances correspond to the midpoints of the population rings around each mill. For each meteorological condition (wind speed and atmospheric stability class) and variable emission rates, the MPTER model indicated the number of days at each receptor where TRS concentrations would exceed the TRS odor threshold of 0.005 mg/m^3.[2] The concentration at any given receptor exceeded the threshold if 98 out of 100 simulations exceeded the threshold.

For each mill and for each hour at each receptor, we estimated the TRS concentration using the 1984 emissions without the Clean Air Act. If the TRS concentration exceeded 0.005 mg/m^3, we then estimated the concentration using the 1984 emissions with the Clean Air Act. Using these concentrations for each receptor, we next estimated the average values for the receptors in each ring. Finally, for each mill and each receptor ring, we estimated the average number of days that had at least one 1-hour average TRS concentration above 0.005 mg/m^3 without the Clean Air Act and the average number of days with at least one 1-hour average TRS concentration above 0.005 mg/m^3 with the Clean Air Act.

Summary air quality profiles for three of the 60 mills are presented in Table 5-2 to show how we aggregated the model's outputs and interpreted them for air quality impact assessments. For example, TRS emissions for Mill 733 would have increased from 150 to 340 tons per year between 1973 and 1984 because of increased production. They actually decreased to 80 tons, however, because of the Clean Air Act. The hypothetical 1984 emissions without the Clean Air Act would have resulted in 44 days that exceeded an odor threshold of 0.005mg/m^3 in ring five (between 5 and 10 miles around the mill). The actual 1984 emissions resulted in only seven days that exceeded the odor threshold in ring five.

Table 5-2.   Average Number of Days with at Least One 1-hour Average
TRS Concentration Greater than 0.005 mg/m^3

Mill / Ring	Households	1973	1984 w/o CAA	1984 w/ CAA
Mill 700				
1	0	111	111	111
2	1,700	107	111	106
3	810	97	111	90
4	4,750	51	111	33
5	15,536	24	77	24
Mill 733				
1	1,290	111	111	111
2	10,630	106	111	97
3	17,900	92	106	59
4	41,790	44	92	34
5	87,610	34	44	7
Mill 811				
1	0	110	110	110
2	660	110	110	92
3	880	110	110	52
4	500	110	110	29
5	2,980	77	86	3

Estimated Air Quality Improvements

The summary results of the air quality simulations without (Scenario
2) and with (Scenario 3) the Clean Air Act emission limits for the 60 kraft
mills showed varied outcomes (Figure 5-4). Without the Clean Air Act
benefits-based standard for TRS, emissions from 39 mills would have
exceeded the odor threshold between 30 and 45 days per year, and TRS
emissions from 31 mills would have exceeded the odor threshold between
20 and 30 days per year. With the Clean Air Act benefits-based standards
for TRS, emissions from only four mills exceed the odor threshold between
20 and 25 days per year, and emissions from none of the mills exceed the
odor threshold over 25 days per year.

Figure 5-4. Number of Days Around 60 Mills with Estimated
Concentrations Greater than 0.005 mg/m³

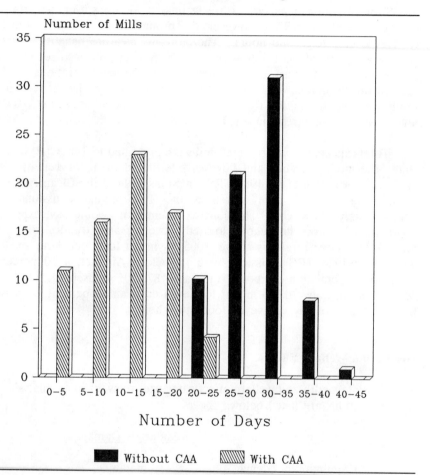

MONETARY BENEFITS

Society values mitigating offensive odors because the reductions
decrease adverse health and welfare effects. Estimation of societal values
in monetary terms requires assessment of the changes in exposure due to
improvements in air quality, calculation of the potential adverse health and
welfare effects, and valuation of these health and welfare effects in
monetary terms.

Exposure

Typically, human exposures are estimated for long averaging times (e.g., one year). For TRS, however, the primary concern is with short-term exposures (e.g., one hour). Therefore, to measure benefits, it was necessary to estimate not only reductions in exposure but also reductions in the number of hours of exposure to different concentrations. We focused on the number of hours and the number of days that households would be able to smell TRS. As discussed in the next section, 0.005 mg/m^3 is the odor threshold for TRS.

Total exposures for three kraft mills are presented in Table 5-2, which shows the model's results and how we interpreted them for benefit-cost analysis. For example, the 1984 TRS emissions without the Clean Air Act at mill 733, would have increased the 1973 actual emissions by approximately 120 percent. These increased emissions would have exposed approximately 87,600 households located between 5 and 10 miles from the mill to TRS concentrations above 0.005 mg/m^3 at least one hour on 44 days. The 1984 TRS emissions with the Clean Air Act, a 50 percent reduction in emissions, exposed the same 87,600 households to TRS concentrations above 0.005 mg/m^3 for seven days, an 85 percent decrease in the number of days above the odor threshold.

Health and Welfare Effects

Possible health and welfare effect categories relevant in an analysis of TRS reductions are listed below:

Health Effects
- mortality
- pulmonary edema
- inflammation/irritation of eyes and respiratory tract

Materials Damage
- paints
- metals

Nuisance
- odor

Community Damage
- property value
- community reputation

At least two conditions must be met if effects data are to be used in the benefit analysis: (1) the TRS effect should preferably be stated in the form of a concentration-response function or at least as a threshold value, and (2) the TRS effect should be amenable to valuation.

Very little information is available on the effects of TRS compounds discharged from kraft pulp mills on human health, animals, vegetation, and materials [NAS. 1979]. Almost all of the information relates only to hydrogen sulfide (H_2S), the predominant TRS compound emitted by kraft pulp mills. This section summarizes information from two EPA reports [EPA. 1979, 1986] on the effects of hydrogen sulfide only. Essentially no information is available on health and welfare effects of the other reduced sulfur compounds (methyl mercaptan, dimethyl sulfide, and dimethyl disulfide) emitted from kraft pulp mills.

Health. Table 5-3 summarizes the effects of hydrogen sulfide at various ambient concentrations. At sufficiently high concentrations, hydrogen sulfide is very toxic to humans. At lower concentrations, it causes pulmonary edema and inflammation of the eyes and the respiratory tract.

Concentrations of TRS as high as 10 mg/m^3 would not likely have been realized near existing kraft mills even without the Clean Air Act. For example, ambient hydrogen sulfide concentrations were measured during a 6-month period in 1961 and 1962 in Lewiston, Idaho. The major contributor of gaseous pollutants was a pulp mill that controlled only the recovery furnace for TRS emissions. The measured levels of hydrogen sulfide were generally less than 0.015 mg/m^3 [EPA. 1979].

EPA performed dispersion modeling studies on typical kraft pulp mills for the purpose of setting the TRS emission guidelines for existing mills. These studies indicated that the maximum ground-level ambient concentration of hydrogen sulfide resulting from a large uncontrolled (1,000 tons/day) kraft pulp mill would be about 10 mg/m^3 (maximum 1-hour average). This level, even though much higher than actually measured at the mill, is still lower than the minimum concentration that causes eye irritation.

Most studies dealing with health effects of kraft pulp mill odors are inconclusive. The studies show that populations in the area of an uncontrolled kraft pulp mill are annoyed by the odor, and that short-term effects (vomiting, headaches, shortness of breath, dizziness) occur in some individuals after prolonged exposure. These effects have been reported to be of a psychosomatic nature, but the evidence in this regard is not conclusive.

Table 5-3.   Health Effects -- Hydrogen Sulfide[a]

Exposure (mg/m^3)	Health Effect
1,440	rapidly fatal (within seconds) -- respiratory center of brain ceases functioning, breathing stops; H_2S also prevents enzymatic activity for use of O_2 by cells.
420	pulmonary edema (short-term exposure)
210 - 350	odor perception lost
70 - 140	inflammation of eyes and respiratory tract (intermittent exposure)
	pulmonary edema possible (prolonged exposure)
20	ocular lesions (4-5 hours of exposure)
	blurred vision (short-term exposure)
	perception of colored rings around light sources (short-term exposure)
14 - 28	irritation of respiratory tract and eyes (long-term exposure)
0.07 - 7.3	equivocal evidence of decreased activity in heme synthesis pathway enzymes (8-hour exposure)
0.05 - 0.01	no adverse effects [Alberta, Canada study]
0.004 - 0.028	odor perception threshold

a. Summarized from EPA's *Health Assessment Document for Hydrogen Sulfide* [1986].

Nuisance. The characteristic odor of kraft mills is principally caused by a mixture of hydrogen sulfide, methyl mercaptan, dimethyl sulfide, and dimethyl disulfide. These sulfides are extremely odorous and may be detectable at concentrations as low as 0.0014 mg/m^3. The odor perception thresholds of these gases vary considerably among individuals; they apparently depend on the age and sex of the individuals and whether they smoke. The odor threshold of hydrogen sulfide varies between 0.004 and 0.028 mg/m^3. The odor becomes more intense as the concentration increases. At very high concentrations (210-350 mg/m^3), the smell is not as pungent, probably due to paralysis of the olfactory nerves.

At the ambient ground-level concentrations likely to occur near an uncontrolled kraft pulp mill, as determined by EPA dispersion estimates, odors would definitely be perceptible. Even a kraft pulp mill that operates with typical emission controls would most likely have a slightly perceptible odor.

Materials Damage -- Paint. Hydrogen sulfide in the atmosphere reacts with paint containing heavy metal salts in the pigment and then dries to form a precipitate which darkens or discolors the surface. Lead, mercury, cobalt, iron, and tin salts result in a gray or black discoloration; cadmium salts result in a yellowish-orange discoloration. The communities of Lewiston, Idaho, and Clarkston, Washington, reported damage to house paint caused by hydrogen sulfide emissions from a kraft pulp mill in the 1960s [EPA. 1979].

Materials Damage -- Metal. Copper and silver can tarnish rapidly in the presence of hydrogen sulfide. Copper that has been exposed to unpolluted air for some time resists attack by hydrogen sulfide, however. Some alloys of gold will tarnish when exposed to hydrogen sulfide. Finally, hydrogen sulfide will attack zinc at room temperature. A zinc sulfide film is formed which prevents further corrosion. At concentrations normally found in the atmosphere and at ambient temperatures, hydrogen sulfide is not corrosive to ferrous metals.

Valuation of Effects

Given the limitations of the data, the only effect considered in the following benefit analysis is odor nuisance. The effects assessment is limited to concentrations exceeding a threshold of 0.005 mg/m^3, the lower end of the odor threshold, rather than a concentration-response function.

At concentrations below the thresholds, no odors are detected, and therefore no effect is said to occur. Above this level, the magnitude of the effect modeled in this study increases in proportion to both the number of persons and the number of days per year that the concentrations are above the thresholds.

Although we conducted an extensive search of the literature and research sources, we could not find any studies that attached a dollar value to reductions of odors from kraft pulp mills.[3] In fact, the only odor valuation study we found was an EPA-funded report valuing reductions in diesel odor in Philadelphia [Lareau and Rae. 1989]. This study forms the basis for the dollar valuation of odor reductions at kraft pulp mills.

Lareau and Rae determined willingness-to-pay from the respondents' relative preferences using a discrete choice model. They found that households would pay between $6 and $8 per year for a reduction of one mild diesel smell per week and between $16 and $21 per year for a reduction of a more intense diesel smell per week. On an odor event basis, they calculated the willingness-to-pay to reduce one exposure to be between $0.11 and $0.40. In addition, they asked the respondents to record the number and intensity of odor contacts they experienced in the week following the survey. On average, the respondents recorded 2.2 contacts per week with mild diesel odor and 4.2 contacts per week with more intense diesel odor. On the basis of this information, Lareau and Rae estimated that the average household in the Philadelphia metropolitan area would be willing to pay approximately $75 annually to completely avoid all diesel odor exposures.

For the purpose of this analysis, we developed three valuation schemes to estimate the monetary benefits of reducing TRS emissions (Table 5-4). Each scheme assigned values based on the percentage of days that at least one 1-hour average TRS concentration fell below the odor threshold (0.005 mg/m^3) due to lower TRS emissions. For each mill and each receptor ring, we multiplied the number of households in the ring by an annual willingness-to-pay value. The annual willingness-to-pay value varied, depending on the percentage of days with reduced TRS odor. For example, households in receptor rings where the days with noticeable TRS odor was reduced between 50 and 75 percent were assumed to be willing to pay $25 per year under the minimum valuation scheme.

Table 5-4.   Valuation Schemes for TRS Reductions ($1984)

Valuation Scheme	Percentage of Days Reduced Below 0.005 mg/m³ Odor Threshold as a Result of Clean Air Act	Willingness To Pay per Household
Minimum	0.00 - 0.50	$ 0
	0.50 - 0.75	$25
	0.75 - 1.00	$50
Mid-Point	0.00 - 0.25	$ 0
	0.25 - 0.50	$25
	0.50 - 0.75	$50
	0.75 - 1.00	$75
Maximum	0.00 - 0.25	$25
	0.25 - 0.50	$50
	0.50 - 0.75	$75
	0.75 - 1.00	$100

Estimated Benefits

Table 5-5 combines the data on exposure changes and values to obtain monetary estimates of benefits for 60 mills and to compare welfare benefits to pollution reduction costs. The three benefits estimates reflect possible combinations of exposure changes and monetary values to give low, mid-point, and high benefit estimates. The cost estimates are the differences in costs between the with and without Clean Air Act scenarios.

On a case-by-case basis, the number of mills that meet the benefit-cost criterion varies significantly between the lower-bound and mid-point assumptions about benefit values and only slightly between the mid-point and upper-bound assumptions. Whereas at the lower-bound value 20 mills meet the criterion, at the higher-bound value 31 mills meet it, and at the mid-point value 29 mills meet it.

Table 5-5. Estimated Benefits and Costs of the Clean Air Act (TRS) for 60 Kraft Pulp Mills ($1984 10^3)

Mill	Change in Days	Annual Benefits Min	Mid	Max	Annual Costs
700	31	$960	$1,470	$2,040	$1,460
702	35	50	70	90	930
703	42	160	240	320	560
704	73	530	800	1,070	890
705	40	350	530	710	1,170
708	36	10	20	30	520
712	36	90	140	180	830
713	40	340	530	720	1,470
716	37	90	140	180	1,190
720	55	140	210	320	1,330
722	51	160	230	310	460
723	39	380	570	750	780
724	43	1,300	1,940	2,600	300
725	52	130	210	300	1,150
726	53	400	600	800	380
729	12	320	580	960	540
731	66	770	1,160	1,540	120
733	31	5,360	9,000	12,960	1,150
734	40	6,020	9,080	12,150	1,020
735	8	830	1,250	1,680	560
736	30	30	60	100	1,450
737	33	290	460	670	1,360
739	45	530	850	1,170	830
742	60	390	590	800	1,020
743	37	1,740	3,030	4,550	2,210
744	68	2,420	3,660	4,900	700
746	77	3,000	4,550	6,110	850
747	35	150	230	310	1,390
748	79	140	210	280	480
751	41	1,250	1,870	2,500	840
754	20	700	1,140	1,580	1,140
759	25	240	420	610	4,460
761	54	1,010	1,510	2,020	1,080
764	49	380	590	840	470
766	54	1,770	2,810	3,930	1,170
770	51	110	180	340	610

(cont.)

Table 5-5. (cont.) Estimated Benefits and Costs of the Clean Air Act (TRS) for 60 Kraft Pulp Mills ($1984 10^3)

Mill	Change in Days	Annual Benefits Min	Mid	Max	Annual Costs
772	56	$320	$480	$640	$1,110
788	49	480	750	1,030	150
791	54	120	170	270	390
795	59	110	160	210	400
797	20	160	250	340	1,290
808	44	220	380	660	1,560
811	48	200	320	450	2,020
822	9	740	1,150	1,630	420
827	74	880	1,330	1,800	200
828	39	210	310	440	230
835	36	140	240	400	1,400
837	56	3,870	5,880	7,910	930
841	45	570	860	1,150	1,570
854	19	30	120	210	1,190
855	28	1,340	2,140	3,000	750
858	33	5,020	7,530	10,060	200
861	34	900	1,410	2,060	610
864	22	270	440	610	200
865	28	5,760	9,120	13,060	980
868	25	150	230	300	370
872	45	110	180	250	1,310
875	52	140	250	360	310
882	55	730	1,120	1,510	900
888	52	420	740	1,110	1,480
	Total	$55,780	$55,720	$86,880	$120,250

Possible Biases in the Benefit Calculations

An obvious source of downward bias in our reported benefit calculations is the simplicity of the air quality modeling. We assumed that the terrain around each mill was flat, even though many mills are located in mountain valleys. As a result, we underestimated the number of hours and days with concentrations above the odor threshold. We could not measure the magnitude of this bias, however.

Another potential bias is that the population distribution around the 60 evaluated mills may not be the same as the population distribution around the 129 study mills for which we had population data. Upon examination of the population distributions, however, we could not see any reason to believe that the population distribution in our sample differed from the study population. As in the case of the ambient analysis in Chapter 4, the 60 mills on average have slightly larger populations around them than the 129 mills with TRS emissions.

FINDINGS

The Clean Air Act benefits-based standard for TRS, as applied to the pulp and paper industry, met the benefit-cost criterion based on the 60 mills in our analysis. As shown in Table 5-6, the benefits exceed the costs based on the mid-point and upper bound benefit estimates. In addition, assuming that the mid-point benefit best characterizes the actual environmental outcome, there are net benefits at approximately 50 percent of the mills.

Table 5-6. Benefits and Costs Attributable to the Clean Air Act Benefit-Based Standard for TRS ($1984 10^6)

Valuation Scheme	Mills with B>C / Mills Analyzed	Total Benefits	Total Costs	Net Benefits
Minimum	20 / 60	$55.7	$55.8	($0.1)
Mid-Point	29 / 60	$86.9	$55.8	$31.1
Maximum	31 / 60	$120.3	$55.8	$64.5

Earlier, we estimated the total costs attributable to the Clean Air Act's benefits-based standard for TRS emissions from pulp mills at $115 million. The average annual TRS control cost attributable to the Clean Air Act is approximately $0.9 million per mill, which is the same as the average annual TRS control costs for the mills in the benefit-cost analysis. Assuming that the 60 mills are representative of the 129 mills, the benefits of TRS reductions attributable to the Clean Air Act are most likely 50 percent greater and possibly 115 percent greater than the costs of TRS control under the Clean Air Act.

This conclusion might be surprising to those not familiar with the establishment of New Source Performance Standards under the Clean Air Act. As explained in Chapter 2, the establishment of designated pollutant standards under the Clean Air Act, particularly for welfare-related pollutants, are benefits-based standards. However, we were surprised that the benefits-based standard for TRS was not that much more successful than the ambient-based standard for TSP in generating net benefits at specific mills. Whereas the TRS benefits-based standard resulted in only one-half of the mills in the sample meeting the benefit-cost criterion, the TSP ambient-based standard resulted in one-third of the mills in the sample meeting the benefit-cost criterion. National standards, although they might meet the net benefits criterion in the aggregate, can fail to meet it in many specific situations if local conditions are not taken into account.

Our finding reinforces the major findings about the technology and ambient approaches. First, a regulatory policy that is based on some measure of environmental results, either ambient or benefits, will be more successful in generating net benefits than a policy that ignores environmental results. Second, really efficient policies (those with many specific situations where the benefits exceed the costs) for reducing environmental risks require pollution mitigation decisions that take into account the number of people who will benefit from pollution reduction decisions, whether they be the size of the recreation market or the number of people exposed to pollutants. Changes in ambient quality per se will not necessarily generate net benefits.

Notes

1. There are offensive odor problems from kraft pulp mills in Jacksonville, Florida as reported by the Associated Press [1990] and in Tacoma, Washington as reported by Tim Eagan [1988] .

2. Monte Carlo simulations generated hourly variation in emission rates for each mill. The simulations were based on a standard deviation of one-third the annual average hourly emission rate.

3. In an environmental management study of the pulp and paper industry prepared for EPA, Putnam, Hayes and Bartlett, Inc. [1984] obliquely estimated a value for eliminating odor emissions from kraft pulp mills. They derived an appropriate value by comparing odor with another aesthetic effect, visibility improvement. The comparison was accomplished by assuming that a unit reduction in the concentration of an odoriferous pollutant is equivalent in value to the same unit reduction in concentration of the most important contributing pollutant for visibility effects, TSP. After some testing, they selected a value which implies that a 1 $\mu g/m^3$ change in TSP concentration is worth $0.26 per person per year. Converting this per microgram value to units which are compatible with the measurement of the odor effect yields an estimated value of $0.23 per household per day for elimination of odors. They believed that this is a conservative (i.e., low) estimate of the actual value of reduction in odor effects.

References

Associated Press. February 7, 1990. "Unsealed Grand Jury Report Hits Industry for Odors in Florida City." Chicago Tribune Co., Chicago, IL.

Brandes, Debra A. (ed.). 1984. *1985 Post's Pulp and Paper Directory.* Miller Freeman Publications, San Francisco, CA.

Eagan, Timothy. April 6, 1988. "Tacoma Journal; On Good Days the Smell Can Hardly Be Noticed." Special to the New York Times Co., New York, NY.

E.H. Pechan and Associates. 1987. "Air Pollution Control Costs for Pulp and Paper Mills." Report to the Office of Policy, Planning, and Evaluation, U.S. Environmental Protection Agency. Springfield, VA.

Lareau, Thomas J., and Douglas A. Rae. 1989. "Valuing WTP for Diesel Odor Reductions: An Application of Contingent Ranking Technique." *Southern Economics Journal*, 55:728-742.

Medynski, Ann L. (ed.). 1973. *Post's 1974 Pulp and Paper Directory.* Miller Freeman Publications, San Francisco, CA.

National Academy of Sciences (NAS). 1979. *Odors from Stationary and Mobile Sources*. National Academy of Sciences, Washington, DC.

Putnam, Hayes and Bartlett, Inc. 1984. "Analysis of Cost Effective Pollution Control Strategies in the Pulp and Paper Industry." Draft report to the Office of Policy, Planning, and Evaluation, U.S. Environmental Protection Agency. Cambridge, MA.

U.S. Environmental Protection Agency (EPA). 1979. "Kraft Pulping: Control of TRS Emissions from Existing Sources." EPA-450/2-78-0036. Office of Air Quality Planning and Standards, Research Triangle Park, NC.

_____. 1980. "User's Guide for MPTER, A Multiple Point Gaussian Dispersion Algorithm with Optional Terrain Adjustment." EPA-600/8-80-016. Research Triangle Park, NC.

_____. 1986. "Health Assessment Document for Hydrogen Sulfide." EPA-600/8-86/026A. External Review Draft, Washington, DC.

Appendix 5-A

KRAFT PULP MILLS EVALUATED IN THIS STUDY

1973 and 1984

The first step in the analysis was to examine the inventory of kraft pulp mills identified in Chapter 4. The 1973 snapshot contains 115 mills producing kraft pulp, and the 1984 snapshot contains 129 kraft pulp mills.

The second step was to identify major sources of emissions at each mill. A review of the unit level National Emission Data System records revealed incomplete data for many mills [EPA. 1984]. To correct the situation to some degree, we assumed that each kraft mill had at least one standard piece of equipment for an identified pulping process. Each mill must have some combination of digesters, evaporators, recovery furnaces, smelt dissolving tanks, and lime kilns. There can be several such pieces of equipment, but there must be a least one of each to produce kraft pulp.

The third step in the analysis was to estimate the pulp capacity for each kraft mill. To compute annual production, we assumed, just as in Chapter 4, that all mills operated 345 days per year.

The fourth step was to identify the pollution control equipment in place for each process and the level of emission control at each kraft mill. We took the information for 1973 from a 1976 EPA inventory [EPA. 1979] and for 1984 assumed that each process had the mandated technology installed or its equivalent unless the plant was reported to be out of compliance.

The fifth step was to determine whether process units in existing mills were in compliance with their emission limits. We determined compliance with TRS emission limits using both EPA records and inquiries with compliance personnel at state air pollution control agencies [Smith. 1986]. We found five non-compliance situations resulting from TRS emissions, but were told that appropriate controls had already been applied and that no additional controls were needed.

1994

The background data for snapshot year 1994 come entirely from the inventory compiled according to the description in Appendix 4-B.

References

Smith, Franklin. 1986. "CDS Quick Look Report." Computer printout from the U.S. Environmental Protection Agency. Washington, DC.

U.S. Environmental Protection Agency. 1979. "Kraft Pulping: Control of TRS Emissions from Existing Sources." EPA-450/2-78-0036. Research Triangle Park, NC.

_____. 1984. "1984 National Emissions Report." Office of Air Quality Planning and Standards, Research Triangle Park, NC.

Appendix 5-B

AIR POLLUTION CONTROL COST ESTIMATION (TRS)

We used process-specific cost functions for each of the components identified in Table 4-B-1. For each of the components, we characterized existing and new source emission reductions for TRS.

Data Sources

We took cost information from an EPA study on reducing TRS emissions [EPA. 1979]. We adjusted the cost data to 1984 dollars with the same approach described in Appendix 4-C.

Cost Functions

The process-specific cost functions derived from the above data source are listed in Table 5-B-1.

We used process-specific cost functions for each of the standard pieces of equipment for the kraft pulping process. Three types of costs -- capital, operation and maintenance, and annual -- are computed in each year for each component in place as appropriate. The Appendix 4-C general engineering cost functions and annualized procedures apply here as well.

Reference

U.S. Environmental Protection Agency. 1979. "Kraft Pulping: Control of TRS Emissions from Existing Sources." EPA-450/2-78-0036. Research Triangle Park, NC.

Table 5-B-1. Total Reduced Sulfur Control Costs[a] ($1984)

Process	Capital		Operating and Maintenance	
	Coefficient	Exponent	Coefficient	Exponent
Lime Kiln	15.872	0.663	0.128	1.0513
Digester and Evaporator	0.722	1.236	0.112	1.0815
Recovery	5.931	0.834	1.516	0.6538

a. Cost equations are in the form $y = a * x^b$, where x is the unit capacity in air-dried tons of pulp per day and y is thousands of dollars.

Appendix 5-C

TRS EMISSIONS

Table 5-C-1. Total TRS Emissions for the Kraft Pulp Mills
Covered in This Study

Mill	TRS Emissions (tons)		
	1973	1984 w/o CAA	1984 w/ CAA
700	210	640	150
702	210	230	50
703	90	140	30
704	60	130	50
705	280	300	60
708	892	41	18
712	220	250	50
713	360	400	90
716	220	390	80
720	530	670	70
722	0	230	20
723	390	360	40
724	550	820	10
725	1,090	3,820	60
726	2,040	2,180	20
729	60	100	60
731	180	250	10
733	150	340	80
734	310	310	60
735	70	90	70
736	700	910	70
737	250	280	70
739	90	240	50
742	2,730	3,270	50
743	1,430	1,430	120
744	1,640	2,180	40
746	4,230	5,450	90
747	340	380	70
748	1,500	1,500	20
751	190	240	40

(cont.)

Table 5-C-1. (cont.) Total TRS Emissions for the Kraft Pulp Mills
Covered in This Study

Mill	TRS Emissions (tons)		
	1973	1984 w/o CAA	1984 w/ CAA
754	210	130	50
759	1,070	1,450	220
761	160	310	60
764	230	510	40
766	700	840	70
770	250	340	30
772	40	590	60
788	370	240	30
791	10	200	20
795	250	280	30
797	240	280	70
808	3,050	3,530	120
811	630	720	90
822	60	90	70
827	490	490	10
828	90	110	10
835	390	380	70
837	980	630	50
841	230	1,120	110
854	410	410	150
855	250	270	100
858	140	170	60
861	240	300	60
864	50	80	40
865	250	220	70
868	90	110	40
872	390	500	50
875	120	140	10
886	170	320	30
888	90	150	10

Appendix 5-D

THE MPTER MODEL

The MPTER computer code (Multiple Point source model with TERrain adjustments) provides a method to estimate air pollutant concentrations from multiple sources in rural environments, and can make optional adjustments for slight terrain variations. The algorithm is based on Gaussian modeling assumptions and incorporates the Pasquill-Gifford dispersion parameter values.

MPTER can estimate concentrations at a maximum of 180 receptors from a maximum of 250 point sources. The model uses Gaussian assumptions and techniques to calculated hourly estimates, considering each hour as a steady state period. Required input information consists of point source and hourly meteorological data. Periods from one hour to one year may be simulated, with all output controlled by the user through selection of options.

Features of the algorithm include:

- An optional terrain adjustment as a function of stability class;
- Inclusion of stack downwash;
- Inclusion of gradual plume rise or final rise only;
- Inclusion or omission of buoyancy-induced dispersion of pollutant at the source using the method of Pasquill;
- Input of anemometer height;
- Input of wind profile power law exponents as functions of stability;
- Optional output of the following information: average concentration over length of record, plus highest five concentrations for each receptor for four averaging times (1, 3, 8, and 24 hours), additional user-selected averaging time.

Reference

U.S. Environmental Protection Agency. 1980. "User's Guide for MPTER, A Multiple Point Gaussian Dispersion Algorithm with Optional Terrain Adjustment." EPA-600/8-80-016. Research Triangle Park, NC.

Chapter 6

COMPLIANCE

Pulp and paper mills receive operating permits from states based on meeting EPA ambient and technology standards. These permits specify, among other things, pollutant discharge limitations. Although most mills are in compliance with these discharge limitations, some mills are not. This chapter analyzes whether bringing these non-complying sources into compliance would result in net benefits. The chapter goes beyond compliance for compliance sake -- i.e., not violating the law -- and asks whether the benefits of reducing health and environmental risks achieved by compliance are commensurate with the costs of compliance.

As in our assessment of the efficiency consequences of alternative regulatory approaches, we will evaluate compliance strategies using the same generic approaches. We will examine technology-based, ambient-based, and benefits-based strategies for dealing with identified non-compliance. After a brief general discussion of the compliance issue, we will evaluate the efficiency gains associated with three compliance strategies for water and then three strategies for air.

MEASURING COMPLIANCE

EPA's approach to the measurement and tracking of the compliance status of pollution sources differs substantially for water and air pollution. Although the water pollution compliance program receives much more monitoring data on each major source, neither the air nor the water compliance data are considered to be entirely accurate.

Water

Of all the environmental programs in the United States, the national water program relies most heavily on self-monitoring and reporting by the sources of pollution. Of the more than 60,000 permittees nationwide, about 6,000 are defined as "major sources" of water pollution, based upon criteria related to environmental impact. All permits for discharges into surface waters of the United States contain requirements for self-monitoring and reporting. The program regulations set forth several requirements: a general requirement for monitoring, the use of standard national Discharge Monitoring Reports for both federal and state-approved programs, a minimum reporting frequency of once per year, and a 3-year recordkeeping requirement. Most major sources must report on a monthly or quarterly basis. Minor sources generally report annually or semi-annually. State governments usually require more frequent reporting.

EPA's Permit Compliance System identifies mills not in compliance with their discharge permits on the basis of filed Discharge Monitoring Reports. Mills are considered to be in significant non-compliance if: (1) their discharges exceed their monthly average permit limits for nontoxic pollutants by 40 percent or for toxic pollutants by 20 percent for any two months in a 6-month period, or (2) their discharges exceed their monthly average permit limits for any four months in a six-month period. For the purpose of this study, compliance is defined as meeting the 30-day monthly average permit limits for BOD and TSS.

Air

Approximately 29,000 major stationary sources of air pollution are regulated in State Implementation Plans, an additional 2,100 new sources are regulated by New Source Performance Standards, and another 1,300 are covered by National Emission Standards for Hazardous Air Pollutants.

Data on air emissions from these stationary sources are much more limited than data on water effluents. Because of the high costs involved, most sources are not required to monitor their air emissions continuously, or to report their emissions to the state or federal government. In addition, pollution control agencies do not frequently inspect them [Pedersen. 1981].

EPA's Compliance Data System provides the only systematic data on compliance status. However, Compliance Data System data are very limited. They do not indicate why a source was in violation or by how much it exceeded its emission standard. Data of this type are obtained only by contacting state pollution control agencies.

Rates of compliance have several significant shortcomings as a single measure of success [Wasserman. 1985]. First, compliance rates are only as good as the statistical adequacy and reliability of the compliance monitoring data used in the calculation. This is particularly troublesome in the air program, which relies heavily on infrequent inspections to provide a single data point on compliance. Because programs usually assume compliance for those sources for which a violation has not been detected, the results of an announced annual inspection may be used to assume year-round compliance. At issue is whether the compliance rate should reflect violations detected for the total universe of regulated sources or just those sources for which sound data are available -- i.e., sources that are inspected or for which self-surveillance data are received.

Second, even a high rate of compliance does not necessarily indicate success from an environmental standpoint, particularly if violators cause great damage to the environment. Similarly, low rates of compliance should not necessarily be interpreted as a program failure if they are the result of improved methods of detecting violations, the application of more stringent standards for defining compliance, or the introduction of new requirements.

WATER COMPLIANCE

EPA's Permit Compliance System reported that 247 major pulp and paper mills filed Discharge Monitoring Reports in 1984. Of these 247 mills, 50 were not complying with their permits. By 1986, the Permit Compliance System reported that 260 mills had filed at least one monthly Discharge Monitoring Report. Of these 260 mills, 61 were not complying with their permits.

The Cost of Clean model (the model used in this study) provides a different measure of compliance than the Permit Compliance System in 1984 for two reasons. First, the Cost of Clean model contains information on 306 of the approximately 335 direct-discharging mills that operated in 1984. Second, the measure of compliance is more stringent. Compliance is determined by comparing a mill's long-term average effluent levels with its 1984 30-day average permit limit. The long-term average effluent levels are based on all the data available in state and federal records and consist of 2 to 26 months of data for each mill. A mill is deemed out of compliance if its long-term average level exceeds its permitted level.

On the basis of data in the Cost of Clean model, 49 mills (15 percent of the mills in Cost of Clean model) were out of compliance in 1984. Twenty-six mills exceeded their permit limits for BOD, eight mills exceeded their limits for TSS, and 16 mills exceeded their permit limits for both BOD and TSS.

A comparison of the Permit Compliance System and Cost of Clean model data for 1984 shows very little overlap (Table 6-1). Eight mills were reported out of compliance by both the Permit Compliance System and the Cost of Clean model for 1984. A comparison of three years of Permit Compliance System data for 1984-1986 and Cost of Clean model data for 1984 shows 15 mills in common as not being in compliance.

Table 6-1. Comparison of Non-Complying Mills Based on the
Permit Compliance System (PCS) and the
Cost of Clean (COC) Model

EPA Region	PCS		COC Model	
	Total Major Mills	Non-complying Mills	Total Mills	Non-complying Mills
1	32	6	48	15
2	20	8	36	4
3	13	0	18	7
4	54	19	67	6
5	68	1	77	7
6	24	13	28	0
7	1	0	2	1
8	0	0	0	0
9	6	0	3	1
10	29	3	27	8
TOTAL	247	50	306	49

Source: EPA, Permit Compliance System and Cost of Clean Model Results.

Technology-Based Compliance

The primary objective of compliance should not be bringing non-complying mills into compliance with their technology-based permits.

Rather, the aim ought to be achieving compliance that results in net benefits.

Our approach for addressing this issue is essentially the same as in Chapter 3 on technology-based regulation. A water quality model, RGDS, is used to simulate changes in water quality resulting from compliance with technology-based standards.

The criterion for water quality violation is more stringent here and in the following analyses than for the water quality benefit analysis in Chapter 3. The analysis in Chapter 3 examined river reaches with only pulp and paper mills and, in a few cases, municipal dischargers. This analysis examines varied river reaches, some of which have many pollutant dischargers. To ensure that pulp and paper mills, the only sources modeled, are not contributing to violations of water quality standards, the acceptable water quality threshold for a major water quality problem is raised from 5.0 to 6.5 mg of dissolved oxygen per liter. In the remaining analyses, any pulp and paper mill discharge that results in an in-stream dissolved oxygen of 5.0 to 6.5 mg/l is considered a minor water quality problem, and any discharge that results in an in-stream dissolved oxygen below 5.0 mg/l is considered a major water quality problem, which is consistent with the previous analysis.

The RGDS model could estimate the water quality conditions resulting from the effluent discharge from only 38 of the 49 non-complying mills (Figure 6-1).[1] The results of the water quality analysis of effluents from non-complying pulp and paper mills show that only one-half of the modeled mills are located on reaches with potential water quality problems. Water quality is a minor problem downstream from 14 mills and a major problem downstream from the remaining seven mills. We judge that water quality downstream from 17 mills would be unaffected by their effluent discharge. Compliance at the 21 problem-causing mills would cost $9.2 million, which is approximately three-quarters of the cost of compliance for all 38 modeled mills.

Enforcement of permit limitations for the 21 mills that are potentially contributing to minor and major water quality problems would reduce effluent discharge. However, reduction of effluent loadings usually results in significant water quality improvements only if current loadings are causing a major violation of water quality standards[2] and the loadings reduction achieved with compliance is a large change in total loadings. Significant water quality improvement is defined as: (1) the dissolved oxygen profile changes to the degree that there is an upward shift in the

Figure 6-1. Water Quality Impacts and Compliance Costs for the
49 Non-Complying Mills Identified by the
Cost of Clean Model (Annual Costs -- $1984 10^6)

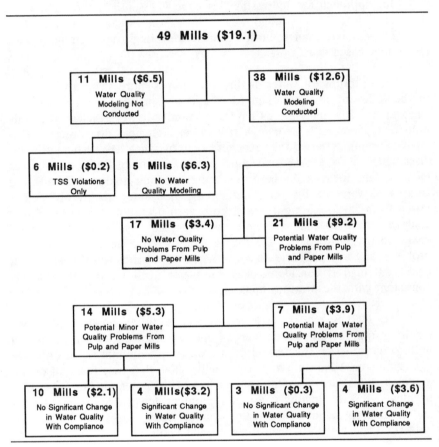

potential use pattern (unusable, boatable, rough fishing, game fishing, or
superior game fishing), or (2) if the use pattern stays the same, 25 percent
or more of the river mileage in that use pattern shifts upwards.[3] The
results of the additional water quality analysis show that enforced
compliance with discharge permits would make a significant change in
water quality at only eight of the 21 mills. Four of the mills are located
on river reaches with major water quality problems, and four mills are
located on river reaches with minor water quality problems. For the other
13 mills, the effluent reductions resulting from compliance with their
discharge permits would not make a significant change in water quality.

Compliance at the eight mills would cost $6.8 million, which is three-quarters of the cost of compliance for the 21 mills potentially contributing or causing water quality problems and is slightly more than one-half of the cost of compliance for all 38 modeled mills.

We applied the benefit-cost analysis used in Chapter 3 to the eight mills where compliance would result in significant water quality improvements.[4]   Chapter 3 compared the economic benefits, primarily enhanced recreation opportunities, with the pollution control costs of achieving water quality improvements.   The lower- and upper-bound benefit estimates for benefits compliance reductions at these eight mills bracket the compliance costs, and thus benefits are assumed to exceed costs (Table 6-2). However, at three of the mills, the compliance costs would exceed the benefits. At these mills, an investment in compliance would not result in net benefits, even though water quality per se would be improved. This is because the user population near these mills is small and the improvement in the quality of the recreation experience would be marginal.

We conclude that complying with the Clean Water Act technology-based standards barely meets the efficiency criterion for the 49 non-complying mills based on our sample of 38 mills. The aggregate benefits from achieving compliance at the 38 mills range from $6 million to $19 million, which results from benefits at only five of the 38 mills.   The aggregate costs from achieving compliance at the 38 mills are approximately $13 million. Assuming that the 38 mills are representative of the 49 mills, then the aggregate benefits could range from $8 million to $25 million, and the aggregate costs would be $19 million. Given that the aggregate benefit estimates bound the aggregate costs of compliance, compliance at the 49 non-complying mills would result in net benefits. However, an interpolated mid-point benefit estimate of $16 million is less that the costs of compliance. This benefit estimate calls into question the value of achieving compliance at all 49 mills and suggests that achieving compliance at only a subset of the 49 mills would result in net benefits.

An obvious question arising from this analysis of the Cost of Clean model of non-complying mills is whether the same general conclusion would result from an analysis of the non-complying mills in the Permit Compliance System. Complete response to this question is not possible because Permit Compliance System data are not available for estimating the non-compliance and compliance loadings from the 50 mills that are out of compliance. The only analysis possible is for the eight non-complying mills in both the Permit Compliance System and Cost of Clean model, and

Table 6-2. Benefit-Cost Analysis of a Technology-Based
Water Compliance Strategy ($1984 10^3)

Mill	Change in Water Quality	Annual Benefits		Annual Costs	Benefits >Costs ?
		Min	Max		
35	R-G*	$2,850	$7,120	$70	yes
73	G-G*	30	50	140	no
144	U-B	670	4,690	90	yes
213	G-G*	120	190	2,970	no
217	R-G*	50	130	2,990	no
252	G-G*	240	400	50	yes
260	U-R	1,840	6,140	500	yes
289	G-G*	90	150	30	yes
Subtotal (5 mills where B>C)		$5,690	$18,500	$740	yes
Total		$5,890	$18,870	$6,840	no

U = Unusable. Dissolved oxygen less than 3.5 mg/l. Major water quality
    problem.
B = Boatable. Dissolved oxygen between 3.5 mg/l and 4.0 mg/l. Major
    water quality problem.
R = Rough fishing. Dissolved oxygen between 4.0 mg/l and 5.0 mg/l.
    Major water quality problem.
G = Game fishing. Dissolved oxygen between 5.0 mg/l and 6.5 mg/l.
    Minor water quality problem.
G* = Superior game fishing. Dissolved oxygen above 6.5 mg/l. No water
    quality problem.

even then the loadings data are from the Cost of Clean model, rather than
Permit Compliance System. The RGDS model could estimate downstream
water quality for only five of the eight mills; the other three mills either
discharge into bays or estuaries or are in violation for only TSS discharge.

Water quality downstream is judged to be unaffected by one mill, to be in minor violation for two mills, and to be in major violation for two mills. As a result of bringing the four mills with minor or major water quality problems into compliance, water quality would not significantly change downstream for two mills and would significantly change downstream for the other two mills. The finding from this analysis of a subset of the Cost of Clean model data, which reflects the Permit Compliance System records on non-compliance, is similar to the finding from the analysis of the full Cost of Clean model data. Approximately one-half of the non-complying mills is causing a water quality problem, and only reducing the discharges of one-quarter of the mills would contribute to a significant improvement in water quality.

Ambient-Based Compliance

Another measure of non-compliance is whether the streams that receive waste discharge from pulp and paper mills are in violation of ambient water quality standards. Ambient standards are attractive in that they most closely represent the social goal of environmental protection -- namely, the reduction in concentration of harmful substances in the ambient environment. They are difficult to use as enforcement standards, however, and are not clearly associated with one source of pollution. Usually, more than one industrial source and many non-point sources (such as agricultural activity) contribute to violations of ambient water quality standards in a river reach.

In spite of the complexity of this issue, we analyzed ambient-based compliance starting with EPA's 1986 "National Water Quality Inventory," usually referred to as the 1986 Inventory [EPA. 1987]. The 1986 Inventory report describes whether existing water quality conditions are adequate to support state-designated use patterns (activity levels). Forty-two states provided information that allowed a determination of attainment with designated uses on 370,000 stream-miles. Although these stream-miles represent only 20 percent of the stream miles in the U.S., they account for approximately 60 percent of the river reaches and 60 percent of the 600,000 stream miles in the Reach File, the basis for RGDS water quality modeling. An overlay of the reaches (approximately 10 miles per reach) in the 1986 Inventory and the directly discharging pulp and paper mills in the Cost of Clean model matched 21 reaches that contained a mill and were not supporting or only partly supporting their designated use pattern. Only in few cases did the 1986 Inventory report identify a pulp and paper mill as the source of the water quality problem, however.

For our analysis, we combined this list of 21 mills from the 1986 Inventory with the 68 mills analyzed in Chapter 3 to construct a working list of 20 potentially polluting mills.[5] We took seven mills from the 1986 Inventory list. Five of these mills contribute to major or minor water quality problems with both low and average summer flows. We took 13 of the 68 mills analyzed in Chapter 3 because they are contributing to major water quality problems. Twelve of the 13 mills contribute to major or minor water quality problems with both low and average summer flow conditions. Approximately twice as many potentially polluting mills are derived from the benefit analysis mill list, which suggests that the 1986 Inventory provides only a partial picture of water quality violations.

The primary aim of achieving ambient water quality compliance should not be the elimination of violations of water quality standards per se. Rather, the aim ought to be the reduction of negative externalities to the extent that the benefits exceed the costs. However, addressing this issue is more difficult than addressing the same issue about technology-based compliance. The identified pulp and paper mills may not be a major or even an important source of the water quality problems described in the 1986 Inventory. In addition, reducing loadings from the mills may or may not make a difference in water quality. However, we are reasonably confident that our approach, the same as the water quality benefit analysis in Chapter 3, identifies those mills where reduced pollutant loadings could result in improvements in water quality.

We started by assessing whether additional wastewater treatment would mitigate the water quality problems (Table 6-3). For 11 mills, none of the three Best Conventional Technology options nor post-aeration of the effluent would alter the dissolved oxygen profile sufficiently to change or to improve significantly the use classification of the receiving waters. For most of these mills, the dilution ratio is very low or the mill discharge is actually greater than average summer flow. This left us with nine mills for which there would be some potential for achieving water quality improvements.

We then analyzed the costs of additional controls and the potential water quality improvements for these nine mills. We found that for four of the nine mills, post-aeration of the effluent at an annual cost of $0.4 million would eliminate major water quality problems.[6] For four of the remaining five mills, meeting the effluent limits of one of the Best Conventional Technology options at an annual cost of $5.9 million would eliminate major water quality problems. For one mill (Mill 296), an

Table 6-3. Benefit-Cost Analysis of an Ambient-Based Water Compliance Strategy ($1984 10^3)

Source	Mill	Low Flow Water Quality	Post Control Water Quality	Effective Pollution Reduction Action	Annual Benefits Min	Max	Annual Costs	Benefits >Costs?
1986	11	U	G	PA	$310	$1,250	$150	yes
Inven-	21	U	-	none[a]	-	-	-	
tory	76	R	G	BCT1	30	80	60	yes
	198	R	G	BCT1	110	270	160	yes
	235	B	G	PA	570	1,900	100	yes
	296	U	R	BCT3	330	1,090	3,350	no
	318	U	G	PA	150	610	110	yes
Benefit	42	U	-	none[b]	-	-	-	
List	43	U	-	none[b]	-	-	-	
	44	U	-	none[c,d]	-	-	-	
	51	R	G	PA	90	220	30	yes
	56	U	-	none[c,d]	-	-	-	
	77	U	-	none[a]	-	-	-	
	79	U	-	none[a]	-	-	-	
	82	U	-	none[a]	-	-	-	
	85	U	-	none[a]	-	-	-	
	192	U	G	BCT2[d]	180	710	540	yes
	216	U	-	none[a]	-	-	-	
	280	U	-	none[a]	-	-	-	
	293	U	G	BCT3[d]	390	1,570	5,100	no
	Subtotal (7 mills where B>C)				$1,440	$5,040	$1,150	yes
	Total				$2,160	$7,700	$9,600	no

PA = Post-Aeration.

BCT = Best Conventional Technology.

a. 1984 loadings are less than the most stringent BCT option. Post-aeration of effluent would not eliminate the water quality problem.

b. Dilution ratio is less than 5.0 and 1984 loadings are less than the most stringent BCT option. Post-aeration of effluent would not eliminate the water quality problem.

c. Plant wastewater flow is greater than average summer-flow. Post-aeration of effluent would not eliminate the water quality problem.

d. BCT option 1 would reduce their pollutant discharge.

investment of \$3.4 million would bring water quality only up to the rough fishing level (dissolved oxygen = 4.0-5.0 mg/l). Only the complete elimination of discharge would eliminate water quality problems for this mill. In summary, our analysis shows that an annual investment of approximately \$6.3 million in pollution control would eliminate major water quality problems for eight of the nine mills. An additional investment of \$3.4 million would reduce, but not eliminate, major water quality problems at the other mill.

We next applied benefit-cost analysis (as used in Chapter 3) to these nine mills. Table 6-3 summarizes the findings of the benefit-cost analysis. In the aggregate, an ambient-based strategy at all nine mills is not justified on the basis of comparing the economic benefits with the pollution control costs. However, the results would support additional effluent reductions at seven mills. The findings reinforce the proposition that the potential for human use as well as water quality changes per se should be a consideration in requiring additional pollution control.

We conclude that complying with an ambient-based compliance strategy does not meet the benefit-cost criterion for the 20 mills that are contributing to water quality problems. An ambient-based compliance strategy would not result in net benefits because the costs of meeting ambient standards would exceed the potential benefits for the nine mills where additional effluent reduction would be possible. An ambient-based strategy would result in a positive benefit-cost outcome, however, at seven of the nine mills. It is interesting to note that none of the nine mills covered by the ambient-based compliance strategy would be covered by the technology-based compliance strategy because they were in compliance with their permit limitations in 1984.

Benefits-Based Compliance

Pursuing a benefits-based compliance strategy would, by design, generate net benefits. If pursued correctly, such a strategy would result in a reasonable balance between the benefits and costs of eliminating water quality violations.

A benefits-based compliance strategy builds on an ambient-based strategy, but a benefits-based strategy incorporates the potential for human use as well as water quality changes per se in designating which mills should pursue additional pollution reductions. As such, it limits the ambient-based compliance strategy to only mills where the benefits of recreation enhancements would exceed the costs of additional pollution reduction.

Our evaluation of the efficiency of a benefits-based compliance strategy began with an inventory of the pulp and paper mills potentially contributing to major water quality problems. We compiled the inventory from our analysis of the technology-based and ambient-based compliance strategies. We took four mills that are not complying with technical standards from Table 6-2 and the nine mills analyzed for water-quality improvements from Table 6-3 to form an inventory of 13 mills (Table 6-4).

Table 6-4. Benefit-Cost Analysis of a Benefits-Based
Water Compliance Strategy ($1984 10^3)

Source	Mill	Change in Water Quality	Annual Benefits Min	Max	Annual Costs	Benefits >Costs ?
COC	35	R-G*	$2,850	$7,120	$ 70	yes
Model	144	U-B	670	4,690	90	yes
List	217	R-G*	50	130	2,990	no
	260	U-R	1,840	6,140	500	yes
1986	11	U-G	310	1,250	150	yes
List	21	R-G	30	80	60	yes
	198	R-G	110	270	160	yes
	235	B-G	570	1,900	100	yes
	296	U-R	330	1,090	3,350	no
	318	U-G	150	610	110	yes
Benefit	51	R-G	90	220	30	yes
List	192	U-G	180	710	540	yes
	293	U-G	390	1,570	5,100	no
Subtotal (10 mills where B>C)			$6,800	$20,200	$1,810	yes
Total (13 mills)			$7,570	$25,780	$13,250	yes

These are mills that potentially contribute to major water quality problems and for which reducing effluent loadings has the potential to improve water quality significantly.

We next applied the benefit-cost framework to the 13 mills where reduction of effluent loadings would result in a significant improvement of water quality (Table 6-4). We eliminated three mills because the costs of pollution reduction would exceed the anticipated economic benefits -- primarily enhanced recreation opportunities. By design, the 10 mills targeted for effluent reductions under a benefits-based water compliance strategy would meet the benefit-cost criterion. Increased annual costs of approximately $1.8 million would produce annual benefits of $6.8-20.2 million.

Findings -- Water

In summary, neither one of the conventional compliance strategies pursued by EPA -- compliance with technology-based permit limitations or compliance with ambient-based effluent reductions -- would produce net benefits (Table 6-5). Both strategies would impose pollution control costs that would exceed the anticipated recreation benefits. The technology-based compliance strategy would be less wasteful of resources, i.e., the resulting recreation benefits would be reasonably proximate to the costs of pollution control.

Table 6-5. Summary of the Benefit-Cost Assessments of Water Compliance Strategies ($1984 10^6)

Compliance Approach	Number of Mills Affected	Annual Benefits	Annual Costs	Benefit-Cost Ratio[a]
Technology-Based[b]	49	$8.0 - $25.0	$19.1	<1.0
Ambient-Based	9	$2.2 - $7.7	$9.7	0.5
Benefits-Based	10	$6.8 - $20.2	$1.8	7.5

a. The ratio is based on a mid-point benefit estimate.
b. Benefit estimates are extrapolated to reflect all the mills that would be affected by the regulatory approach.

The preferable approach for reducing environmental risks from an economic efficiency perspective is a benefits-based compliance strategy.[7] This strategy would require only a subset of the mills that are non-complying with technology-based permits or potentially causing major water quality problems to reduce their effluent discharge. The subset of mills would include only those individual mills where the benefits exceed the costs. Our analysis of that subset of mills shows a very positive benefit-cost ratio. The strategy is so successful because it starts with environmental problems and limits the imposition of pollutant reduction to rivers where there is the potential for a significant improvement in water quality and a sufficiently large recreation demand for the water quality improvements. The Clean Water Act does not currently encourage either EPA or the states to pursue such a policy.

AIR COMPLIANCE

EPA's Compliance Data Systems reported on 383 pulp mills in the four major production Standard Industrial Classification categories in 1984. The Cost of Clean model for the same year contains 157 mills, all listed in the Compliance Data System inventory. The 157 mills accounted for 94 percent of pulp production in 1984 [Brandes. 1984]. By 1986, the Compliance Data System reported on 396 mills.

A combination of Compliance Data System reports and state records provides perspective on the compliance status of pulp and paper mills. In 1984, CDS indicated that 38 mills (10 percent) were not in compliance for TSP. Based on state records, however, only 17 mills were exceeding their emission limits. The other 21 mills were either complying or not complying for administrative reasons. In 1986, CDS reported on 396 mills. Of these, 45 mills (11 percent of the total) were in not in compliance. Based on state records, however, only 19 mills were exceeding their emission limits. As in 1984, the other 26 mills either were complying or were not complying for administrative reasons.

We did not attempt to calculate an independent measure of compliance with the Cost of Clean model, as we did for water permit compliance. The task was too difficult because there are approximately 1,600 air emission points at the 157 mills (rather than only one, or perhaps two, water effluent sources at each mill). Also, the source monitoring data are sparse and not subject to quality control procedures.

Technology-Based Compliance

Although air operating permits were at some time in the past based on eliminating violations of ambient-based standards, we suspect that they are now very similar to technology-based permits. Mills listed in the Compliance Data System are judged to be out of compliance because pollution reduction equipment is not being operated to achieve an agreed-upon emission limitation. Consequently, we treat the violations in the Compliance Data System as violations of technology-based standards. Just as in the case of water permit compliance, the primary aim of compliance should not be bringing non-complying mills into compliance with their operating permits. Rather, the aim ought to be achieving compliance that results in net benefits.

Our approach for addressing the issue is essentially the same as in Chapter 4. An air quality model, MCP, is used to simulate changes in air quality resulting from compliance with operating permits. We could model air quality conditions with and without compliance for only seven of the 17 mills out of compliance for excess TSP emissions, however. For the other 10 mills, we could not obtain data about the nature of the excess TSP emission violation and/or the cost of correcting the violations from EPA or state agencies. The application of the MCP model is indicative of air quality conditions around the seven non-complying sources because they account for more than 50 percent of total TSP emissions in their respective counties.

The MCP modeling indicates that excess emissions from the seven mills could contribute to an air quality problem based on a threshold of 40 $\mu g/m^3$ for adverse health effects (Figure 6-2). The non-compliance emissions added to background emissions around all seven mills exceeded 40 $\mu g/m^3$. On the basis of the initial air quality analysis, we cannot eliminate any of the seven mills as potential candidates for enforcement actions.

Enforcement of permit limits at the seven mills that are potentially contributing to air quality problems would reduce TSP emissions, but would not necessarily result in a significant improvement in air quality. To refine the analysis, we first separated potential air quality problems, as we did for the water quality analysis, into minor potential problems (i.e., ambient concentrations between 40 and 50 $\mu g/m^3$ around mills) and into major potential problems (i.e., ambient concentrations above 50 $\mu g/m^3$) in order to follow the analysis pattern for water quality compliance problems. As a result, five mills were categorized as contributing to minor air quality problems, and two mills were categorized as contributing to major air quality problems.

Figure 6-2. Air Quality Impacts and Compliance Costs for the 17 Non-Complying Mills Identified by the Compliance Data System (Annual Costs -- $1,984 10^6)

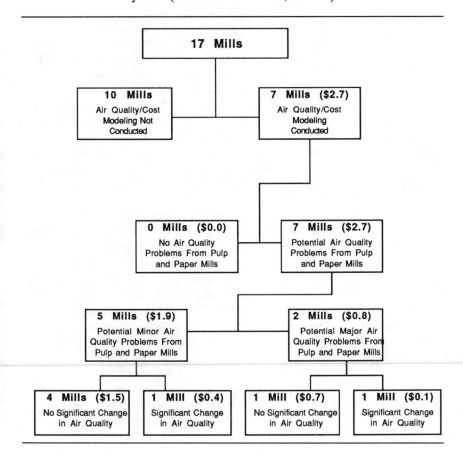

We then separated air quality changes resulting from compliance into significant and non-significant improvements. A "significant" change for the purpose of compliance analysis is a modeled reduction in TSP concentrations of 0.1 $\mu g/m^3$ or greater. The 0.1 $\mu g/m^3$ change reflects the findings in Chapter 4 that all of the 60 mills showed a modeled improvement in TSP concentrations of 0.1 $\mu g/m^3$ or more between the with and without the Clean Air Act scenarios. (The median change at the 60 mills was 1.5 $\mu g/m^3$.) The MCP modeling indicates that air quality would significantly improve as a result of compliance at two of the seven mills. Compliance at these two mills would reduce emissions by more than 100 tons a year (Table 6-6). Compliance at the five mills with non-significant compliance would result in emission reductions of less than 100 tons per year.

The results of the additional air quality/exposure analysis at only two mills show that emission reductions resulting from compliance would make a significant change in air quality at a cost of $0.5 million. For the other five mills, the emission reductions resulting from compliance would not make a significant change in air quality, but would result in expenditures of $2.2 million. Given that there are only seven mills in the analysis, we cannot easily generalize from the results of the analysis. However, we are surprised that a compliance strategy focused on achieving "significant" air quality improvements at two mills would require only 20 percent of the full compliance expenditures. Achieving universal compliance appears to be expensive, and the resulting emission reductions appear to be small.

We next applied the benefit-cost analysis used in Chapter 4 to the seven non-compliance mills, with an emphasis on the two mills where compliance would result in significant air quality improvements. Chapter 4 compared the economic benefits, primarily reductions in adverse health effects, with the pollution control costs of achieving air quality improvements. Both the lower- and upper-bound benefit estimates for these seven mills are less than the compliance costs (Table 6-6). However, the benefits of compliance at one mill, given the mid-point estimate, exceed the compliance costs. At this mill, an investment in compliance would result in net benefits.

We conclude that complying with a technology-based compliance strategy does not meet the benefit-cost criterion for the seven non-complying mills. The aggregate benefits range from less than $0.1 million to $0.9 million, whereas the aggregate costs are approximately $2.7 million.

Ambient-Based Compliance

Another measure of non-compliance is whether the counties in which the mills are located are not attaining ambient air quality standards. As with ambient water quality standards, violations of ambient air quality standards are not clearly associated with one source of pollution. Usually, more than one industrial source and many area sources (such as shopping centers) contribute to violations of ambient air quality standards.

Table 6-6. Benefit-Cost Analysis of a Technology-Based Air Compliance Strategy ($1984 10^3)

Mill	Air Quality	Tons Reduced	Popula- tion	Annual Benefits			Annual Costs	Benefits >Costs?
				Min	Mid	Max		
708	minor/s	340	10,500	$0	$10	$10	$370	no
717	major/ns	30	8,500	0	0	10	680	no
772	minor/ns	20	20,600	0	0	0	560	no
890	minor/ns	20	542,200	10	70	150	200	no
821	minor/ns	70	16,100	0	0	10	310	no
879	minor/ns	60	19,100	0	10	30	400	no
891	major/s	130	68,900	40	340	730	140	yes
Subtotal (1 mill with B>C)		130	68,900	$40	$350	$740	$510	yes
Total (7 mills)		670	685,900	$50	$420	$940	$2,660	no

s = Significant change in air quality (0.1 $\mu g/m^3$ or greater).
ns = Non-significant change in air quality (less than 0.1 $\mu g/m^3$).

In spite of the complexity of the issue, we identified a set of mills that are potentially contributing to air quality violations based on EPA's "National Air Quality Inventory" [EPA. 1984, 1985, 1986]. The National Air Quality Inventory (the Air Quality Inventory) lists the attainment status of all counties for each criteria air pollutant and indicates whether all or part of the county is in non-attainment. All states are required to report on attainment status (contrary to the Water Quality Inventory), so the Air Quality Inventory provides complete coverage for the U.S. We examined the latest Inventories, covering the years 1984-1986, and found 314 counties or county equivalents with non-attainment areas in 1984, 290 counties or county equivalents in 1985, and 268 counties or county equivalents in 1986. We compared the non-attainment counties over the

3-year period with the mills in the Cost of Clean inventory. Of the 157 mills, only 42 mills were located in non-attainment counties. The EPA Inventories do not identify the sources within these counties causing air quality problems, however.

To test whether these 42 mills were contributing to air quality problems, we focused on the 15 mills in counties exceeding the primary (health-based) TSP ambient standard. The other 27 mills are located in counties exceeding the secondary (welfare-based) TSP ambient standard. We used the MCP model to simulate ambient concentrations around 12 of the 15 mills at their 1984 production levels. (We did not have data to simulate concentrations around the other three mills.) Assuming that an ambient concentration of 60 $\mu g/m^3$ or greater indicates a potential ambient air quality problem, we retained six mills for additional analysis. The other six mills are assumed not to contribute to the non-attainment problem in their counties.

We expanded our list of potential air quality problems by examining the approximately 110 mills (including the 13 already under consideration) around which we initially simulated TSP ambient concentrations for the ambient air quality analysis in Chapter 4. Again, assuming that an ambient concentration of 60 $\mu g/m^3$ or greater indicates a potential ambient air quality problem, we added seven mills to the list of potential contributors to air quality problems. The selection of these mills is partially verified by the fact that four of the seven mills are in counties exceeding the secondary TSP ambient standard.

We next assessed whether emission reductions were possible at the 13 mills and whether the reductions would result in significant improvements in air quality. For six mills, five taken from the inventory, no additional emission reductions were possible based on the data available in the National Emissions Discharge System. (We think that emission reductions would be possible, if detailed engineering studies were made at these mills.) For six mills, Reasonably Available Control Technology could be installed and would achieve TSP emission reductions greater than 50 tons per year (Table 6-7). For one mill, an emission offset technology could be installed and would achieve TSP reductions greater than 50 tons per year. The imposition of additional pollution control technology at these seven mills would reduce emissions by 7,400 tons at an annual cost of $6.9 million. These emission reductions would result in significant improvements in ambient air quality at only four of the seven mills, however. The imposition of additional pollution control technology at these four mills would cost $5.2 million. If the objective is to secure a significant improvement in ambient air quality, the additional investment of $1.7 million at the three other mills would not be warranted.

We next applied the benefit-cost framework used in Chapter 4 to the seven mills where additional reductions in emissions would be possible. Table 6-7 summarizes the findings of the benefit-cost analysis. In aggregate, an ambient-based compliance strategy at the seven mills is not justified based on a comparison of the mid-point economic benefits with the pollution control costs at the seven mills. The positive benefit-cost findings with the maximum benefits estimate are distorted by the benefits at one mill. The results more realistically support imposing additional emission reductions at two of the seven mills. The findings reinforce the proposition that human exposure to pollutants as well as air quality changes per se should be a consideration in requiring additional pollution control.

Table 6-7. Benefit-Cost Analysis of an Ambient-Based
Air Compliance Strategy ($1984 10^3)

Mill	Air Quality	Tons Reduced	Popula- tion	Annual Benefits			Annual Costs	Benefits >Costs?
				Min	Mid	Max		
Inventory								
709	major/ns	300	22,000	$0	$30	$60	$730	no
728	major/s	2,800	164,800	1,400	3,400	7,700	1,220	yes
754	-	-	-	-	-	-	-	-
831	-	-	-	-	-	-	-	-
842	-	-	-	-	-	-	-	-
845	-	-	-	-	-	-	-	-
865	-	-	-	-	-	-	-	-
MCP								
736	major/s	1,440	6,500	180	510	1,120	660	yes
791	major/s	920	12,600	170	480	1,050	1,410	no
807	major/ns	330	36,000	10	20	50	570	no
811	major/s	1,500	15,400	50	150	330	1,900	no
829	-	-	-	-	-	-	-	-
835	major/ns	110	21,500	10	20	50	380	no
Subtotal (2 mills with B>C)		4,240	171,300	$1,580	$3,910	$8,820	$1,880	yes
Total		7,400	278,800	$1,820	$4,610	$10,300	$6,870	no

s = Significant change in air quality (0.1 $\mu g/m^3$ or greater).
ns = Non-significant change in air quality (less than 0.1 $\mu g/m^3$).

We tentatively conclude that compliance with the Clean Air Act's ambient-based standards does not meet the benefit-cost criterion for the seven mills whose emissions are contributing to violations of ambient air quality standards and whose emissions could be reduced from current control levels. Most likely, an ambient-based compliance strategy would not result in net benefits because the costs of meeting ambient standards would exceed the benefits. Implementation of an ambient-based compliance strategy that imposes additional pollution reductions only on those mills contributing to air quality problems would result in some improvement in air quality, however. In fact, the ambient-based compliance strategy examined in this effort would reduce emissions from seven mills not covered by the technology-based compliance strategy.

Benefits-Based Compliance

Pursuing a benefits-based compliance strategy would, by design, generate net benefits. If pursued correctly, such a strategy would result in a reasonable balance between the benefits and costs of eliminating air quality violations. A benefits-based compliance strategy builds on an ambient-based strategy. A benefits-based strategy incorporates the potential for human health protection as well as air quality changes per se in designating which mills should invest in additional emission reductions. As such, it limits the ambient-based compliance strategy to only those mills where the benefits of pollution reduction would exceed the costs.

Our evaluation of the efficiency of a benefits-based compliance strategy began with an inventory of pulp mills contributing to potentially major air quality problems. We compiled the inventory from our analysis of the technology-based and ambient-based compliance strategies. We took two mills that are not in compliance with their operating permits from Table 6-6 and seven mills that are potentially violating ambient air quality standards from Table 6-7 to form an inventory of nine mills (Table 6-8). These are mills that potentially contribute to major air quality problems. For most of these mills, reducing emissions would significantly improve air quality.

We next applied the benefit-cost framework to the nine mills. We eliminated six mills from the inventory of mills under consideration because the costs of pollution reduction would exceed the anticipated benefits. By design, the three mills targeted for emission reductions under a benefits-based air compliance strategy would meet the benefit-cost criterion. Increased annual costs of approximately $2.0 million would produce annual benefits of $1.6 - $9.6 million.

Table 6-8. Benefit-Cost Analysis of a Benefits-Based
Air Compliance Strategy ($1984 10^6)

Mill	Air Quality	Tons Reduced	Popula-tion	Annual Benefits Min	Mid	Max	Annual Costs	Benefits >Costs?
CDS								
708	minor/s	340	10,500	$0	$10	$10	$370	no
891	major/s	130	68,900	40	340	730	140	yes
1984-86 Inventory								
709	major/ns	300	22,000	0	30	60	730	no
MCP								
728	major/s	2,800	164,800	1,400	3,400	7,700	1,220	yes
736	major/s	1,440	6,500	180	510	1,120	660	yes
791	major/s	920	12,600	170	480	1,050	1,410	no
807	major/ns	330	36,000	10	20	50	570	no
811	major/s	1,500	15,400	50	150	330	1,900	no
835	major/ns	110	21,500	10	20	50	380	no
Subtotal (3 mills where B > C)		6,590	240,200	$1,610	$4,250	$9,550	$2,020	yes
Total (9 mills)		7,670	358,200	$1,850	$4,960	$11,110	$7,380	no

s = Significant change in air quality (0.1 $\mu g/m^3$ or greater).
ns = Non-significant change in air quality (less than 0.1 $\mu g/m^3$).

Findings -- Air

In summary, neither one of the conventional compliance strategies pursued by EPA, compliance with technology-based permit limitations or compliance with ambient-based emission reductions, would produce net benefits (Table 6-9). Both strategies would impose pollution control costs that exceed the anticipated health benefits. The ambient-based compliance strategy would be less wasteful of resources in that the resulting benefits would be reasonably proximate to the costs of pollution control.

Table 6-9. Summary of the Benefit-Cost Assessments of
Air Pollution Compliance Strategies ($1984 10^6)

Compliance Approach	Number of Mills Affected	Annual Benefits	Annual Costs	Benefit-Cost Ratio[a]
Technology-Based	7	$0.1 - $0.9	$2.7	0.1
Ambient-Based	7	$1.8 - $10.3	$6.9	0.7
Benefits-Based	3	$1.6 - $9.6	$2.0	2.0

a. The ratio is based on a mid-point benefit estimate.

The preferable approach for reducing environmental risks from an economic efficiency perspective is a benefits-based compliance strategy.[8] This strategy would require only a subset of mills that are non-complying with technology-based permits or potentially causing major air pollution problems to reduce their emissions. The subset of mills would include only those individual mills where benefits exceed the costs. Our analysis of that subset of mills shows a positive benefit-cost ratio. The strategy is successful because it starts with environmental problems and limits the imposition of additional pollutant reduction to areas where the population exposed to pollution is sufficiently large. However, the strategy is not as successful as the benefits-based water compliance strategy in generating net benefits because of the high incremental costs of reducing air emissions. The Clean Air Act does not currently encourage EPA or the states to pursue such a policy.

FINDINGS

The findings about alternative compliance strategies generally make a strong argument for a benefits approach for reducing environmental risks. The benefits-based compliance strategies for both water and air are the only strategies that would result in net benefits.

In addition, the technology-based strategy for water compliance, but not for air compliance, might also result in net benefits. The technology-based strategy for water compliance, in contrast to air compliance, would result in many mill-specific, positive benefit-cost outcomes. Rather than making a case for the technology approach, this finding reinforces the impression that technology-based standards have prevented pollutant reductions needed to meet ambient water quality standards. Apparently, compliance with technology-based permit limits, rather than elimination of environmental problems, has become the implicit measure of success of the water pollution control program.

Notes

1. The model could not simulate water quality conditions for five non-complying mills because they discharge into bays or estuaries. Nor could it simulate conditions for another six mills because they violated only TSS permit limits. An informed evaluation of the water quality conditions downstream of three of the five non-complying mills not modeled is that water quality is satisfactory (greater than 6.5 mg of dissolved oxygen per liter) and that effluent reductions achieved by compliance would not significantly change water quality conditions. Two of the mills are located on the estuarine portions of the Delaware and Columbia Rivers where the advective dilution ratios are high (in excess of 300). The other three mills are located on bays, harbors, or straits where there is sufficient far-field dilution to mask any pollutant impacts.

2. The modeled dissolved oxygen is less than 5.0 mg/l for some portion of the downstream reach.

3. An example of the latter is that some mileage would be in the 5.0-6.5 mg/l dissolved oxygen category with 1984 non-compliance loadings and the remaining mileage would be in the 6.5-7.0 mg/l dissolved oxygen category. The reach would be classified as a minor water quality problem. Under the compliance simulation, the mileage in the 5.0-6.5 mg/l dissolved oxygen category would decrease by 25 percent or more, and the equivalent mileage would appear in the 6.5-7.0 mg/l or the greater than 7.0 mg/l dissolved oxygen categories. The change would be characterized as a significant improvement in a minor water quality problem.

4. The only difference in the benefit-cost evaluation in this chapter compared to Chapter 3 is to ignore as a "significant" change in water quality any within use category change that affects less than 25 percent of the stream mileage in that use category.

5. To test whether the 21 mills from the 1986 Inventory were actually contributing to water quality problems, we used the RGDS model to simulate water quality profiles similar to those in Chapter 3 for low- and

average summer-flow conditions. The RGDS simulation for low-flow conditions, which are the basis for setting discharge limits more stringent than EPA's effluent guidelines, indicates that seven mills are causing major water quality problems, seven mills are contributing to minor water quality problems, and five mills do not contribute to water quality problems. Two mills could not be modeled. Under average summer-flow conditions, which the 1986 Inventory reflects, only 2 mills are causing major water quality problems, 7 mills are contributing to minor water quality problems, and 10 mills do not contribute to water quality problems. Again, two mills could not be modeled. Even the two mills that are causing water quality problems according to our criterion may not be causing a significant problem because the stream mileage below 5.0 mg/l dissolved oxygen is very short in both situations (0.3 and 0.1 mile), and those areas could be the established mixing zones for diluting pollutant discharges.

We expanded this list of mills with potential water quality violationss by examining the 68 mills analyzed in Chapter 3. These mills, if they are causing a water quality problem according to the RGDS simulation, are likely violations because they are the only significant sources on the streams. Seventeen mills in compliance with their permit conditions are contributing to a major water quality problem under low-flow conditions according to RGDS water quality simulations. Four of these mills are already on the list based on the 1986 Inventory, so we are only adding 13 more mills to our list. Under average summer-flow conditions, only six mills cause major water quality problems, six mills contribute to minor water quality problems, and one mill does not contribute to water quality problems. The six mills that cause major water quality problems according to the RGDS simulations are located on reaches with limited assimilative capacity.

6. We use the term "eliminate" water quality problems to refer to raising water quality at least to impaired game fishing levels (dissolved oxygen = 5.0-6.5 mg/l).

7. This approach, although appealing to economists, would not be termed "fair" by affected entities or elected officials.

8. This approach, although appealing to economists, would not be termed "fair" by affected entities or elected officials.

References

Brandes, Debra A. (ed.). 19845. *1985 Post's Pulp and Paper Directory*. Miller Freeman Publications, San Francisco, CA.

Pedersen, William F. 1981. "Why the Clean Air Act Works Badly." *University of Pennsylvania Law Review*, 129:1059-1109.

U.S. Environmental Protection Agency (EPA). 1984. "Maps Depicting onattainment Areas Pursuant to Section 107 of the Clean Air Act - 1984." EPA-450/2-84-006. Office of Air Quality Planning and Standards, Research Triangle Park, NC.

_____. 1985. "Maps Depicting Nonattainment Areas Pursuant to Section 107 of the Clean Air Act - 1985." EPA-450/2-85-006. Office of Air Quality Planning and Standards, Research Triangle Park, NC.

_____. 1986. "Maps Depicting Nonattainment Areas Pursuant to Section 107 of the Clean Air Act - 1986." EPA-450/2-86-006. Office of Air Quality Planning and Standards, Research Triangle Park, NC.

_____. 1987. "National Water Quality Inventory: 1986 Report to Congress." EPA 440/4-87-008. Office of Water, Washington, DC.

Wasserman, Cheryl. 1985. "Improving the Efficiency and Effectiveness of Compliance Monitoring and Enforcement of Environmental Policies. United States: A National Review." Draft report to the Organization for Economic Cooperation and Development, Environment Directorate. Paris, France.

Chapter 7

GROWTH

An important issue remains to be addressed about conventional pollutants (BOD and TSS), criteria pollutants (TSP and SO_2), and designated pollutants (TRS). How much more pollution mitigation is needed to protect the environment in view of the growth in capacity, and would this additional control result in net benefits?

Ambient water and air quality conditions and Congressional concern support the need to examine the adequacy of the Clean Water and Clean Air Acts in the context of projected industrial growth. First, evidence exists that discharges from pulp and paper mills are contributing to violations of ambient standards. The 1986 National Water Quality Inventory Report and the 1984 - 1986 National Air Quality Inventory Reports document violations of ambient quality standards in the vicinity of pulp and paper mills. Nor are the state reports necessarily complete if the RGDS water and MCP air modeling of current discharges are taken as reasonable estimates of actual ambient quality. The modeling results indicate unreported violations of standards resulting from discharges of pulp and paper mills.

Second, Congress, in the Clean Water Act and Clean Air Act Amendments of 1977, recognized the general necessity for more stringent pollution control on existing sources. The Clean Water Act, in particular, called for regulations more stringent than Best Practicable Technology. Although complex and somewhat vague, the Clean Water Act prescribed "cost tests," rather than water quality considerations, as the basis for Best Conventional Technology regulation [Baum. 1983]. The costs and performance levels of municipal wastewater treatment plants were to be used to establish benchmarks for determining reasonable levels of control for all industry. EPA's final Best Conventional Technology regulation established a benchmark so stringent that the pulp and paper industry would not have to comply with the regulation [FR. 1986(a)]. The Congressionally prescribed basis for the regulation, comparison of municipal wastewater treatment plant costs with industry pollution control costs, precluded EPA from considering water quality impacts or even the more traditional technology/affordability considerations (the basis for Best Practicable Technology regulations) as the basis for more stringent regulation of water effluents.

In a similar spirit, the Clean Air Act requires existing sources to install Reasonably Available Control Technology in non-attainment areas. EPA has defined Reasonably Available Control Technology as:

> devices, systems, process modifications, or other apparatus or techniques, the application of which will permit attainment of the emission limitations ... [provided that these restrictions are] not intended, and shall not be construed, to require or encourage State agencies to adopt such emission limitations without due consideration of (1) the necessity of imposing such emission limitations in order to attain and maintain a national standard, (2) the social and economic impact of such emission limitations, and (3) alternative means of providing for attainment and maintenance of such national standard [CFR. 1985].

Anticipated future increases in capacity in the pulp and paper industry, which we estimated in Appendix A, support the need to examine EPA's policy for mitigating the environmental consequences of economic expansion. We estimated that the annual capacity for pulp production and for paper and paperboard would increase by 8,100 tons (15 percent) and by 9,800 tons (15 percent), respectively, in the 1984-1994 time period (Table 7-1)[1]. This 1984-1994 increase is estimated to occur primarily in the South and in the Midwest. For the purposes of this report, we estimated that approximately 60 percent of the expansion would occur at existing mills; the remaining 40 percent would occur at new mills. The 20 new mills are divided between 14 direct and six indirect water dischargers, and six of these 20 new mills would use kraft pulping, four would use groundwood pulping, and 10 would use secondary fibers.

In spite of the requirement to install more stringent pollution control equipment (New Source Performance Standards) at existing mills with significant capacity expansions and at new mills, water effluents and air emissions are estimated to increase based on current Clean Water Act and Clean Air Act policies (Table 7-1). Conventional pollutants from direct dischargers are estimated to increase by 30,000 tons per year of BOD (15 percent increase) and by 40,000 tons per year of TSS (20 percent increase) between 1984 and 1995. Emissions of criteria pollutants are estimated to increase by 10,000 tons per year of TSP (6 percent) and by 80,000 tons per year of SO_2 (16 percent) between 1984 and 1994.

Designated pollutants for this industry are estimated to increase by 1,000 tons per year of TRS (15 percent) between 1984 and 1994.

Table 7-1. Pulp, Paper, and Paperboard Capacity and Water and Air Pollutant Loadings Included in This Study -- 1973, 1984, and 1994

Year		
Capacity (10^6 tons)		
	Pulp	Paper and Paperboard
1973	46.6	59.1
1984	54.5	67.3
1994	62.6[a]	77.1[a]

	Estimated Pollutant Loadings (10^3 tons)				
	Water		Air		
	BOD	TSS	TSP	TRS	SO_2
1973	700	680	350	60	780
1984	200	190	160	7	710
1994	230	230	170	8	790

a. See Note #1.

CONVENTIONAL WATER POLLUTANTS

This section analyzes the benefits and costs of three approaches that could be pursued to reduce conventional water pollutants during the 1984-1994 period covered by this study. The three approaches are similar to the three historical approaches -- technology, ambient, and benefits -- that structured our evaluation of programs implemented by EPA between 1973 and 1984. The three approaches are configured, however, to conform to realistic options that EPA might pursue in limiting the future discharge of water pollutants. They are: (1) a technology-forcing approach, (2) an ambient antidegradation approach, and (3) a benefits approach. To date, EPA has rejected a technology-forcing approach for existing mills and does not pursue its stated antidegradation approach. EPA has never considered a benefits approach because the Clean Water Act does not require EPA to balance the environmental results of its water regulations against the costs imposed on society. For all three approaches, we assess whether the additional pollution mitigation associated with the approach would result in net benefits.

Technology-Forcing Approach

The Clean Water Act clearly encourages the application of pollution control technology without consideration of potential environmental results. The Clean Water Act Amendments of 1972 established minimum treatment requirements, and the Amendments of 1977 required EPA to develop additional regulations for conventional industrial pollutants. The technology-forcing approach evaluated here is based on EPA's deliberations about Best Conventional Technology [EPA. 1986]. EPA identified four technology options that are capable of removing significant amounts of conventional effluents. EPA's Best Conventional Technology Option 1 -- in-plant production process controls -- is the basis for our technology-forcing approach.

The Best Conventional Technology effluent limitations (pounds of pollutants per ton of production) for each subcategory were applied to the directly discharging 306 pulp and paper mills in the Cost of Clean model inventory at their 1994 capacity levels. For all mills with permit conditions less stringent than the Best Conventional Technology Option 1 limits, the Best Conventional Technology limits became the new permit conditions. For all mills with permit conditions more stringent than the Best Conventional Technology Option 1 limits, the current permit conditions remain the binding effluent constraint.

Implementation of a technology-forcing approach would affect 176 mills out of 306 mills. In particular, it would make permit limits more stringent at 33 of the 60 mills that are estimated to expand their paper and paperboard capacity between 1984 and 1994. This policy would reduce effluent loadings from 230,000 tons to 170,000 tons of BOD (a 25 percent decrease) and from 230,000 tons to 160,000 tons of TSS (a 30 percent decrease). The 176 mills would meet the more stringent limitations at an annual cost of approximately $100 million. Implementation of a technology-forcing approach would increase the annual water pollution control costs of the industry by eight percent.

To assess whether this policy would result in reducing environmental risks, we started by examining its water quality implications. The general implications of such a policy can be simulated with the RGDS water quality model used in Chapter 3. The application of the RGDS model indicates where reductions in effluents have the potential for achieving water quality improvements. The RGDS model could estimate potential water quality conditions resulting from effluent discharge for only 151 of the 176 mills that would be affected by the technology-forcing effluent limitations (Figure 7-1). The model could not simulate water quality conditions for the other 25 mills, located primarily in the West because

Figure 7-1. Water Quality Impacts and Pollution Reduction Costs for the 176 Mills Affected by a 1994 Technology-Forcing Water Policy (Annual Costs -- $1984 10^6)

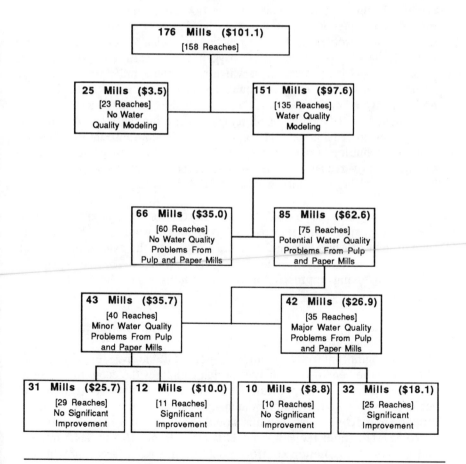

they discharge into bays or estuaries. Simulation of such situations is technically feasible, but requires information and technical effort beyond the scope of this study.

Based on model estimates, we judged that water quality downstream from 66 mills (60 reaches) would be unaffected by effluent discharges. The dissolved oxygen in all segments of the potentially affected reaches would exceed 6.5 mg/l under low-flow stream conditions. We judged that water quality would be a minor problem downstream from 43 mills and a major problem downstream from 42 mills.

Imposition of the technology-forcing effluent limitations at 44 of the 85 mills contributing to minor and major water quality problems would significantly improve water quality as defined in Chapter 6 (Figure 7-1).[2] For the 32 mills on 25 reaches that do not meet game fishing standards, the effluent reductions would make significant improvements in major water quality problems. For 19 reaches on which 26 of these mills are located, the change in dissolved oxygen would shift the potential use pattern. And for six reaches on which six of these mills are located, the change in dissolved oxygen would enhance the current use pattern; however, five of these six reaches are classified as unusable, even after the implementation of technology-forcing effluent limitations. Apparently, more stringent reductions in effluent discharges than those imposed by this minimal technology-forcing policy are necessary to improve water quality for those five reaches. The annual costs of this technology-forcing approach for these 32 mills would be $18.1 million.

For the other group of 12 mills on 11 reaches, the effluent reductions would make significant improvements in minor water quality problems (impairment of game fishing). For the 11 reaches, the change in dissolved oxygen would enhance game fishing recreation. The costs of the technology-forcing approach for these 12 mills would be $10.0 million annually.

We next applied the benefit-cost framework developed in Chapter 3 to the combined group of 44 mills where a technology-forcing policy would result in significant water quality changes. Table 7-2 summarizes the findings of the benefit-cost analysis, which compares the economic benefits -- enhanced recreation opportunities -- with the pollution control costs of achieving water quality improvements.[3] Implementing the technology-forcing approach at these 44 mills would result in a favorable benefit-cost outcome. The water recreation benefits range from $16.3 to $44.7 million, and the costs of pollution control to achieve this result are approximately

Table 7-2. Benefit-Cost Analysis for the 44 Mills Affected by a Technology-Forcing Water Approach ($1984 10^3)

Mill	Change in Water Quality	Annual Benefits Min	Annual Benefits Max	Annual Costs[a]	Benefits >Costs ?
24}	R-G*	$80	$190	$700	No
25}	R-G*	80	190	570	No
27	U-R	310	1,040	260	Yes
28	R-R$_s$	80	210	100	Yes
29	R-G	270	670	400	Yes
35	R-G*	2,850	7,120	170	Yes
52	U-G*	450	1,800	130	Yes
54	U-U$_s$	110	640	1,250	No
74	R-G	170	480	700	No
81	U-G*	1,550	6,210	840	Yes
88	U-U$_s$	40	250	3,070	No
105	G-G$_s$	100	200	1,200	No
107	G-G$_s$	40	60	70	No
111	U-G*$_s$	810	3,220	330	Yes
126	B-G*	190	620	530	Yes
152	U-U$_s$	10	30	2,250	No
155	U-B	20	140	670	No
159	R-G	680	1,650	160	Yes
167	R-G*	90	230	580	No
172	U-U$_s$	40	230	320	No
179	G-G$_s$	440	740	130	Yes
181	G-G$_s$	80	130	20	Yes
192	U-B	40	310	300	Yes
198	R-G	110	270	250	Yes
200}	R-G	90	220	790	No
204}	R-G	90	220	520	No
205}	R-G	90	220	30	Yes
211}	R-G	90	220	590	No
219	B-R	610	2,290	330	Yes
224	G-G$_s$	90	160	20	Yes
228	R-G*$_s$	1,350	3,380	30	Yes
252	G-G$_s$	140	240	1,470	No
262}	G-G$_s$	1,660	2,770	610	Yes
263}	G-G$_s$	1,610	2,680	90	Yes
278	G-G$_s$	160	260	1,640	No
289	G-G*$_s$	90	150	660	No
296	U-U$_s$	40	220	360	No
297	G-G$_s$	20	40	10	Yes
306	G-G$_s$	10	20	4,080	No
320	B-G*$_s$	690	2,290	260	Yes
336}	B-G*	200	670	1,180	No
338}	B-G*	200	670	80	Yes
339}	B-G*	200	670	80	Yes
357}	B-G*	200	670	130	Yes
Total		$16,260	$44,690	$28,080	

a. Based on 1984 capacity.

Abbreviations:

U, B, R = Major water quality problem (dissolved oxygen < 5.0 mg/l).

G = Minor water quality problem (dissolved oxygen 5.0-6.5 mg/l).

G* = No water quality problem (dissolved oxygen > 6.5 mg/l).

s (subscript) = Significant improvement, but use category remains the same.

} = Indicates mills located on same reach. Population divided equally.

$28.1 million. A closer examination of the findings reveals, however, that the upper-bound estimates of the benefits of water quality improvements would exceed the costs of pollution control at only 24 of the 44 mills. At the other 20 mills, the costs of the technology-forcing approach would exceed the benefits.

An obvious conclusion emerges from this evaluation of a technology-forcing approach. Extrapolating from our evaluation of effluent reductions at 151 mills, this policy would not result in net benefits in the aggregate. The effluent reductions resulting from this policy would significantly improve water quality downstream for only 51 of the 176 mills; the reductions would not improve water quality downstream from the other 125 mills. Economic benefits would occur at only those 48 mills, and the benefits would range from $18.0 to $49.0 million. The costs of this policy would be approximately $100 million. The costs would exceed the discernible water quality improvements and measurable economic benefits.

Antidegradation Approach

The Clean Water Act recognizes the importance of an antidegradation approach in its goal to "restore and maintain the chemical, physical, and biological integrity of the nation's waters" (33 U.S.C. Section 101). Although Congress has never explicitly ratified an antidegradation approach, EPA's predecessor, the Federal Water Pollution Control Administration, adopted an antidegradation approach in 1968, and EPA reaffirmed the approach in 1983 [Putter and Jackson. 1986]. An antidegradation approach prevents new and existing sources that expand from degrading ambient water quality, whereas a technology-forcing approach requires new and existing sources to install more stringent effluent reduction measures.

The antidegradation approach evaluated here focuses only on existing sources that are estimated to expand production between 1984 and 1994 and seeks to constrain their total effluent loadings to the 1984 level in spite of increasing production. This policy draws on three of the technology options that EPA identified in its Best Conventional Technology policy determination to reduce the estimated increase in total loadings.

Our analysis of an antidegradation approach began with a matching of the expanding mills and the available Best Conventional Technology options. We took the estimated total BOD loadings in 1994 and ascertained what effluent limitation would be necessary to keep them at their 1984 level. For 49 mills, the effluent limitations resulting from one of three Best Conventional Technology options were sufficient to keep total loadings at their 1984 level. Best Conventional Technology Option 1 limitations were sufficient at 23 mills, Best Conventional Technology Option 2 limitations were sufficient at 17 mills, and Best Conventional Technology Option 3 limitations were sufficient at nine mills. For the other 11 mills, none of the Best Conventional Technology options was found to be sufficient to maintain total loadings at their 1984 level. These mills are currently meeting permit limits more stringent than any Best Conventional Technology effluent limitations.

Implementation of the antidegradation approach would affect 49 out of 60 existing directly discharging mills that expand production. The 49 mills would meet the more stringent limitations at an annual cost of $160 million. By 1994, the existing mills that expand production would have spent approximately $60 million just to stay in compliance with their current permit limits. The antidegradation approach would almost triple their expenditures on pollution control by 1994, assuming that none of the 11 mills was required to meet the most stringent Best Conventional Technology effluent limitations or to change its production processes.

An antidegradation approach not only would constrain total effluent loadings at the 49 mills, but also would lower aggregate BOD loadings from all 60 mills below their 1984 level. It would impose effluent limitations significantly more stringent than necessary to maintain aggregate 1984 loadings. The aggregated BOD loadings from the 60 mills were approximately 50,000 tons per year in 1984 and were estimated to be 65,000 tons per year in 1994. With the antidegradation approach in place, aggregate loadings from the 60 mills are estimated to be 35,000 tons per year. The antidegradation approach would reduce aggregate 1984 loadings from the 60 mills by approximately one-third.

To assess whether this approach would result in reducing environmental risks, we started by examining its water quality implications. The general implications of such an approach can be simulated with the RGDS water quality model. The application of RGDS indicates how the increased loadings from the 49 mills, in the absence of this approach, would affect ambient water quality and how it would improve with this approach. Given that this approach results in total loadings equal to or less than 1984 loadings, the potential for water quality improvements is significant.

Actually the RGDS model could estimate potential water quality conditions resulting from effluent discharge for only 43 of the 49 mills (Figure 7-2). As before, the model could not simulate water quality conditions for the six mills that discharge into bays or estuaries. Simulation of such situations is technically feasible, but requires information and technical effort beyond the scope of this study.

Figure 7-2. Water Quality Impacts and Pollution Reduction Costs for the 60 Mills Potentially Affected by a 1994 Antidegradation Policy (Annual Costs -- $1984 10^6)

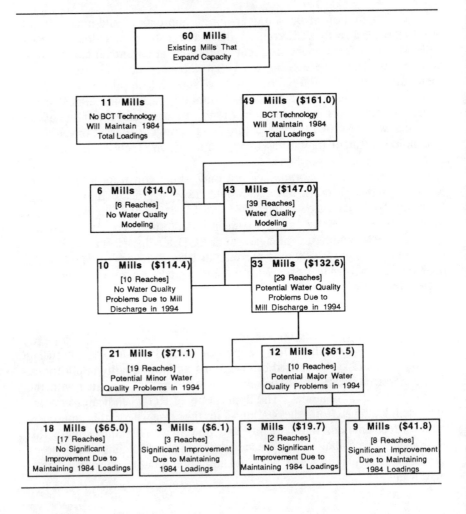

Based on the model estimates, we judged that water quality downstream from 10 expanding mills (10 reaches) would be unaffected by the increased effluent discharge. The dissolved oxygen in all segments of the potentially affected reaches would exceed 6.5 mg/l. We judged that water quality would be a minor problem downstream from 21 expanding mills and a major problem downstream from 12 expanding mills.

The results of the RGDS water quality analysis indicate that the antidegradation effluent limitations would significantly improve water quality at only 12 mills (Figure 7-2). For one group of nine mills on eight reaches, the effluent reductions would make significant improvements in major water quality problems. For seven reaches on which eight of these mills are located, the change in dissolved oxygen would shift the potential use pattern. And for one reach on which one of these mills is located, the change in dissolved oxygen would enhance the current use pattern. This one reach is classified as unusable with 1984 and 1994 effluent loadings. Apparently, a more stringent effluent limitation than the one currently in place is necessary to improve water quality. The costs of meeting this policy for those nine mills would be $41.8 million annually.

For the other group of three mills on three reaches, the effluent reductions would make significant improvements in minor water quality problems. For the three reaches on which the three mills are located, the change in dissolved oxygen would enhance game fishing recreation. The costs of meeting this policy for those three mills would be $6.1 million annually.

We next applied the benefit-cost framework developed in Chapter 3 to the 12 mills where an antidegradation approach would result in significant water quality changes. Implementing this policy even at the 12 mills with significant water quality changes would not result in a favorable benefit-cost outcome (Table 7-3). The water recreation benefits range from $2.3 million to $5.9 million, and the costs of pollution abatement to achieve these results are approximately $47.9 million. A closer examination of the findings reveals that the benefits of water quality improvements would exceed the costs of pollution control at two mills, however. These situations are where the mills would be required to meet the least stringent Best Conventional Technology option at a cost of less than $0.5 million per mill.

Table 7-3. Benefit-Cost Analysis for the 12 Mills Affected by a
1994 Antidegradation Water Approach ($1984 10^3)

Mill	Anticipated 1994 Water Quality	Effective Pollution Control	Post Control Water Quality	Annual Benefits Min	Max	Annual Costs[a]	Benefits >Costs ?
24	R	BCT2	G*	$80	$190	$11,700	No
26	R	BCT3	G	350	1,000	14,400	No
29	R	BCT1	G	270	670	500	Yes
30	R	BCT3	G	270	670	7,300	No
74	R	BCT1	G	170	480	1,400	No
105	G	BCT1	G_s	100	200	1,300	No
155	U	BCT1	B	20	140	1,800	No
159	R	BCT1	G	680	1,650	300	Yes
198	R	BCT2	G	110	270	4,000	No
252	G	BCT1	G_s	140	240	1,600	No
296	U	BCT1	U_s	40	220	510	No
316	G	BCT2	G_s	100	200	3,200	No
Total				$2,330	$5,930	$47,910	

a. Based on estimated 1994 capacity.

U, B, R = Major water quality problem (dissolved oxygen < 5.0 mg/1).
G = Minor water quality problem (dissolved oxygen 5.0-6.5 mg/1).
G* = No water quality problem (dissolved oxygen > 6.5 mg/1).
S (subscript) = Significant improvement, but use category remains the same.

An obvious conclusion emerges from this evaluation of an anti-degradation approach. Extrapolating from our evaluation of effluent reductions at 43 mills, this approach would not result in net benefits in the aggregate. The effluent reductions resulting from this policy would significantly improve water quality downstream from only 14 of the 49 mills; the reductions would not improve water quality downstream from the other 35 mills. Economic benefits from this policy would occur at only those 14 mills, and the benefits would range from $2.7 million to $6.9 million. The costs of this policy, $161.0 million, exceed the discernible water quality improvements and measurable economic benefits.

Benefits Approach

A consistent result emerges from the efficiency evaluations of both the technology-forcing and antidegradation approaches. Uniform implementation of both approaches would not generate net benefits, but implementation of these approaches at a subset of the mills would. This result suggests that we turn to the third generic regulatory approach, a benefits approach, as the preferable option for mitigating the adverse environmental effects of conventional water pollution in the future. A benefits approach would result, by design, in net benefits. If consistently pursued, such an approach would result in a reasonable balance between the benefits and costs of eliminating violations of ambient water quality standards.

A benefits approach builds upon ambient quality conditions and the physical potential for improvements in water quality as a result of pollutant reduction. Its unique feature is that it incorporates the potential for human use as well as water quality improvements in designating which mills should reduce their effluent discharges. As such, it limits concern for ambient improvements only to these mills where the benefits of pollution reduction would exceed the control costs.

Our evaluation of the efficiency consequences of a benefits approach began with an inventory of the pulp and paper mills potentially causing major water quality problems. We compiled the inventory from our analysis of the technology-forcing and antidegradation approaches. We took 24 non-growth mills subject to the technology-forcing approach from Table 7-2 and 10 mills subject to the antidegradation approach from Table 7-3 to form an inventory of 34 mills. These are mills that potentially cause major water quality problems and for which reducing effluent loadings from these mills would significantly (as defined earlier) improve water quality.

We next applied the benefit-cost framework to the 34 mills where reduction of effluent loadings would result in a significant improvement in water quality (Table 7-4). We eliminated 18 mills because the control costs would exceed the anticipated benefits, which are primarily enhanced recreation opportunities. Thus, only 16 mills would be targeted for effluent reductions under a benefits-based policy. With annual costs of $4.3 million and potential benefits of $10.4 million to $32.4 million, the benefit-cost ratio of this policy would be very favorable, being in the range of five to six.

Table 7-4. Benefit-Cost Analysis for the 34 Mills Potentially Affected by a 1994 Benefits Approach ($1984 10^3)

Production Status	Mill	Change in Water Quality	Annual Benefits Min	Max	Annual Costs[a]	Benefits >Costs ?
No Increase	25	R-G*	$80	$190	$570	No
	27	U-R	310	1,040	260	Yes
	28	R-R$_s$	80	210	100	Yes
	35	R-G*	2,850	7,120	170	Yes
	52	U-G*	450	1,800	130	Yes
	54	U-U$_s$	110	640	1,250	No
	81	U-G*	1,550	6,210	840	Yes
	88	U-U$_s$	40	250	3,070	No
	111	U-G*	810	3,220	330	Yes
	126	B-G*	190	620	530	Yes
	152	U-U$_s$	10	30	2,250	No
	167	R-G*	90	230	580	No
	172	U-U$_s$	40	230	320	No
	192	U-B	40	310	300	Yes
	200}	R-G	90	220	790	No
	204}	R-G	90	220	520	No
	205}	R-G	90	220	30	Yes
	211}	R-G	90	220	590	No
	219	B-R	610	2,290	330	Yes
	228	R-G*	1,350	3,380	30	Yes
	320	B-G*	690	2,290	260	Yes
	336}	B-G*	200	670	1,180	No
	339}	B-G*	200	670	80	Yes
	357}	B-G*	200	670	130	Yes
Increase	24	R-G*	$80	$190	$11,700	No
	26	R-G	350	1,000	14,400	No
	29}	R-G	270	670	500	Yes
	30}	R-G	270	670	7,300	No
	74	R-G	170	480	1,400	No
	155	U-B	20	140	1,800	No
	159	R-G	680	1,650	300	Yes
	198	R-G	110	270	4,000	No
	296	U-U$_s$	40	220	510	No
	338	B-G*	200	670	3,200	No
Subtotal (16 mills with B>C)			$10,370	$32,370	$ 4,320	Yes
Total (34 mills)			$12,750	$40,120	$59,750	No

a. Based on estimated 1994 capacity.
Abbreviations:
     U, B, R = Major water quality problem (dissolved oxygen < 5.0 mg/l).
          G = Minor water quality problem (dissolved oxygen 5.0-6.5 mg/l).
         G* = No water quality problem (dissolved oxygen < 6.5 mg/l).
  s (subscript) = Significant improvement, but use category remains the same.
         } = Indicates mills located on same reach. Population divided equally.

Findings -- Water

In summary, the conventional approaches that EPA has traditionally pursued to minimize the environmental risk of pollution resulting from capacity expansion in the pulp and paper industry would not result in a reasonable balance between recreation benefits and pollution reduction costs (Table 7-5). Moreover, a technology-forcing approach, the less costly and less wasteful in a benefit-cost sense, would constrain effluent increases at only one-half of the mills that increase their capacity, which would increase the total effluent discharged.

Table 7-5. Summary Comparison of 1984 - 1994 Regulatory Approaches for Conventional Water Pollutants ($1984 10^6)

Approach	Number of Mills Affected	Annual Benefits[a]	Annual Costs[b]	Benefit-Cost Ratio
Technology-Forcing	176	$18.0 - $49.0	$101.1	0.3
Antidegradation	49	$ 2.7 - $ 6.9	$161.0	<0.1
Benefits	16	$10.4 - $32.4	$4.3	5.3

a. Benefit estimates are extrapolated to reflect all the mills that would be affected by the regulatory approach.
b. Based on estimated 1994 capacity.

The preferable approach for reducing environmental risks is a benefits approach. This approach would require only a subset of the mills subject to a technology-forcing or antidegradation approach to reduce their effluent discharges. The subset of mills would include only those individual mills where the benefits exceed costs. Our analysis of the subset shows a very positive benefit-cost ratio. The policy is so successful because it starts with environmental problems and targets additional pollution reduction to rivers where there is potential for significant improvement in water quality and a sufficiently large recreation demand for water quality improvements. The Clean Water Act with its current emphasis on meeting water quality standards per se (1987 Amendments) does not encourage either EPA or the states to pursue such a policy, however.

CRITERIA AIR POLLUTANTS -- TSP

This section analyzes potential policies for reducing future environmental risks from criteria air pollutants.[4] We estimate the benefits and costs of three approaches to reduce TSP emissions during the 1984-1994 period covered by this study. The three approaches are similar to the three historical approaches -- technology, ambient, and benefits -- that structured our evaluation of programs between 1973 and 1984. However, the three approaches are configured to conform to realistic options that EPA might pursue in limiting future discharge of air pollutants. They are: (1) a technology-forcing approach, (2) a nondegradation approach, and (3) a benefits approach. To date, EPA encourages a technology-forcing approach for existing mills in only those areas of the country that are exceeding the TSP ambient standard and pursues a nondegradation approach only in pristine areas of the country. EPA has never considered a benefits approach because the Clean Air Act has not required EPA to balance the anticipated environmental results of its air program (except in a few cases, such as TRS controls and lead in fuel) against the costs imposed on society. For all three approaches, we assess whether additional pollutant mitigation would result in net benefits.

Our analysis of air policies is more limited and more complex than the analysis of water policies. The air policy analysis is more limited because EPA is not required by statute to develop uniform technology options similar to Best Conventional Technology for existing mills. As a result, the information available on pollution control options is much more limited. In addition, we did not model the air quality implications of emission reductions for as many mills as we did for the water analysis. Consequently, we could not describe the air quality implications of uniform technology policies in nearly as many situations as we could for water quality.

The air policy analysis is more complex because there are several sources of TSP emissions within any one mill. For example, an integrated bleached kraft mill could have as many as 10 distinct emission points. These emissions are released directly to the environment, rather than collected into a single (or perhaps two) discharge points as is the practice for water pollution control.

Technology-Forcing Approach

The Clean Air Act does not encourage the uniform application of pollution control technology on existing sources. The Clean Air Act Amendments of 1970 only require that the emissions from existing sources be reduced to the degree necessary to meet ambient air quality standards. The Clean Air Act Amendments of 1977, however, require that emissions from existing sources in non-attainment areas (areas in violation of ambient air quality standards) install Reasonably Available Control Technology.

The technology-forcing approach evaluated here is based on EPA's definition of Reasonably Available Control Technology for power boilers and process sources at kraft mills [FR. 1978, 1986(b)]. Only those units whose TSP emissions would change 50 tons per year or more as a result of installing Reasonably Available Control Technology are included in the analyses, however.

To perform this analysis, we first separated boiler controls from process controls. For power boilers, we estimated the costs of replacing all low-efficiency cyclones (less than 95 percent control) with electrostatic precipitators that were 95 percent efficient in controlling TSP. Any boiler with at least 95 percent TSP control was assumed to be at the Reasonably Available Control Technology levels already. Calculations of emission changes by upgrading low-efficiency cyclones to precipitators were straightforward. Current operating and maintenance costs of controls were subtracted from the net annual costs of the new equipment to estimate incremental control costs of achieving Reasonably Available Control Technology.

For process controls, we decided to estimate the additional emission controls necessary for all kraft mill process sources to meet a level of control at least one notch above economic recovery. Thus, all process sources controlling to economic recovery levels or below were forced to bring their controls to Reasonably Available Control Technology levels. These levels were established as 99 percent efficient precipitators for recovery furnaces, 95 percent efficient scrubbers for smelt tanks, and 99 percent efficient precipitators for lime kilns. Costs for reaching these control levels were estimated using the same approach used in Chapter 4.

Implementation of a technology-forcing approach based on Reasonably Available Control Technology would affect 59 out of the 163 pulp-producing mills operating in 1994. In particular, it would reduce emissions at 24 of the 54 existing mills that expand their pulp capacity between 1984 and 1994. A technology-forcing approach would reduce aggregate TSP emissions in 1994 from an anticipated 170,000 tons to 140,000 tons, a 20 percent decrease. The 59 mills could achieve this emission reduction at an annual cost of approximately $50 million, which would increase the estimated 1994 annual TSP pollution control costs by 30 percent.

In order to assess whether a technology-forcing approach would result in net benefits, we first examined air quality implications. The general implications can be simulated with the MCP air quality model used in Chapter 4. This application of the MCP model indicates where reductions in emissions would secure significant improvements in air quality.

Actually, the MCP model could estimate potential air quality concentrations resulting from TSP emissions for only 42 of the 59 mills that would be affected by the technology-forcing emission limitations (Figure 7-3). The model could not simulate air quality concentrations for the other 17 mills because meteorological data have not yet been incorporated into the MCP for all areas of the country. The MCP modeling indicates that 1994 emissions from 37 mills could contribute to air quality problems because the ambient concentrations around the mills would exceed 40 $\mu g/m^3$. The other five mills are judged not to contribute to air quality problems because the ambient concentrations around the mills would not exceed 40 $\mu g/m^3$.

Meeting technology-forcing emission limitations at the 37 mills that are potentially contributing to air quality problems would reduce TSP emissions, but would not necessarily result in a "significant" improvement in air quality. To refine the analysis, we first separated potential air quality problems, as we did for the air compliance analysis, into minor potential problems (ambient concentrations between 40 $\mu g/m^3$ and 50 $\mu g/m^3$) and major potential problems (ambient concentrations above 50 $\mu g/m^3$). As a result, 20 mills are categorized as contributing to minor air quality problems, and 17 mills are categorized as contributing to major air quality problems.

We next separated air quality changes resulting from the imposition of the technology-forcing approach into significant and non-significant improvements. A "significant" change for future regulatory action is a reduction in TSP concentration of 1.0 $\mu g/m^3$ or greater within a 1-mile radius of a mill. The hurdle for significance is greater than for the compliance analysis because we are seeking to identify emission reductions

Figure 7-3. Air Quality Impacts and Pollution Reduction Costs for the 59 Mills Affected by a 1994 Technology-Forcing Policy for TSP (Annual Costs -- $1984 10^6)

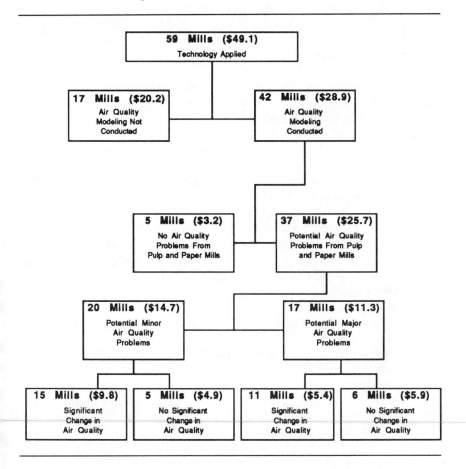

that would be comparable to those achieved as a result of the Clean Air Act. The 1.0 $\mu g/m^3$ change reflects the findings in Chapter 4 that two-thirds of the mills showed a decrease in TSP concentration of 1.0 $\mu g/m^3$ or more between the with and without the Clean Air Act scenarios.[5] The MCP modeling indicates that emission reductions at 11 mills would result in a significant emission reduction. Of these 11 mills, six mills are emitting TSP that could cause major air quality problems, and five mills are emitting TSP that could cause minor air quality problems. The emission reductions at these 11 mills would average 1,200 tons per mill, and over one-half of the mills would reduce emissions by at least 900 tons. The average cost per ton reduced would be $800. Conversely, at the 26 mills where emission reductions would not result in a significant change in air quality, the average ton reduction per mill would be only 400 tons, and only one-quarter of the mills would reduce emissions by more than 500 tons. The average cost per ton reduced would be $1,400.

We next applied the benefit-cost framework developed in Chapter 4 to the combined group of 11 mills where a technology-forcing policy would result in significant improvements in air quality. That chapter compared the economic benefits, primarily avoidance of adverse health impacts, with the pollution reduction costs. In the aggregate, implementing a technology-forcing approach at the 11 mills would probably not result in a positive benefit-cost comparison (Table 7-6). The monetary value of the reductions in adverse health and welfare effects would range from $2.2 million to $12.7 million, with a mid-point estimate of $5.7 million, and the costs of air pollution reduction to achieve this result would be approximately $10.9 million. A closer examination of the findings, moreover, reveals that the monetary benefits of air quality improvements would exceed the costs of pollution reduction at only five of the 11 mills. At the other six mills, the costs of this technology-forcing approach would exceed the benefits. Three out of the five mills that meet the benefit-cost criterion are mills contributing to major air quality problems.

An obvious conclusion emerges from our evaluation of a technology-forcing approach. In the aggregate, this approach would not result in net benefits, extrapolating from our evaluation of emission reductions at 42 mills. Emission reductions from this policy would significantly improve air quality around only 15 of the 59 mills; reductions would not improve air quality around the other 44 mills. Economic benefits from this policy would occur at only those 15 mills, and the benefits would range from $3.0 million to $17.0 million. The costs of this policy would be $50.0 million. Thus, the costs would exceed the discernible air quality improvements and measurable economic benefits.

Nondegradation Approach

The Clean Air Act recognizes the importance of a nondegradation approach, more accurately described as a "no significant deterioration" approach, in its goal "to protect and enhance the quality of the Nation's resources..." [42 U.S.C. Section 7401(b)(1)]. The evolution of this

Table 7-6. Benefit-Cost Analysis for the 11 Mills Affected by a
1994 Technology-Forcing Approach -- TSP ($1984 10^3)

Mill	Air Quality	Tons Reduced	Popula-tion	Annual Benefits Min	Mid	Max	Annual Costs[a]	Benefits >Costs?
728	major/s[b]	2,800	164,800	$1,400	$3,400	$7,700	$1,220	Yes
736	major/s	1,440	6,500	180	510	1,120	660	Yes
776	major/s	660	9,700	0	0	10	450	No
791	major/s	920	12,600	170	480	1,050	1,410	No
811	major/s	1,500	15,400	50	150	330	1,900	No
862	major/s	120	40,100	80	210	450	320	Yes
708	minor/s	390	10,500	0	10	20	660	No
724	minor/s	330	79,800	90	260	580	570	Yes
733	minor/s	280	513,600	160	460	1,000	1,450	No
751	minor/s	2,000	132,000	50	150	330	450	No
797	minor/s	3,200	11,100	10	30	70	1,800	No
Total		13,640	996,100	$2,190	$5,660	$12,660	$10,890	

a. Based on estimated 1994 capacity.
b. s = Significant improvement in air quality (0.1 $\mu g/m^3$ or greater).

approach began with the discussion of air quality standards in the mid-1960s and was approved in the hearings preceding the Clean Air Act of 1970 [Grad. 1985]. The legislative history suggests that state implementation plans should not allow areas with clean air to deteriorate. A nondegradation approach focuses on preventing new and expanding sources of pollution from worsening ambient air quality, whereas a technology-forcing approach requires new and expanding sources to install more stringent emission reduction measures.

The nondegradation approach evaluated here focuses only on those mills that will increase TSP emissions between 1984 and 1994 as a result of mill expansions. These future emission increases would become the targets for obtaining "offsetting" emission reductions. We could assess the general implications of such an approach based on our estimates of capacity expansions that increase TSP emissions. We estimated that 54 existing mills would expand their pulping capacity and, as a result, would increase TSP emissions between 1984 and 1994.

We were not able to develop a computer routine for selecting the least-cost set of control options to reduce the emission increases, so this selection was done manually. In most cases, there was at least one uncontrolled TSP source that could be controlled enough to keep expanding mills at 1984 emission levels. Only at three mills did more than one source have to be controlled to meet the 1984 emission level constraint.

Because emission reductions for only 20 mills were possible with this method, another analysis was performed to see if maximum available controls applied to all coal- and wood-fired boilers and all kraft processes would reduce TSP emissions enough to offset new growth in emissions. Four additional mills were able to control TSP emissions enough through these maximum controls to offset their expected emission increases in 1994. The remaining 30 mills were already at or near maximum available control levels in 1984, so the emission constraint was not achievable. However, at nine mills, the application of Reasonably Available Control Technology could result in some emission reductions, although not enough to constrain emissions to the 1984 level.

In sum, a nondegradation approach would offset emission increases at 33 of the 54 mills that are projected to expand their production capacities between 1984 and 1994. The nondegradation approach would maintain or actually decrease 1994 emissions to 1984 emission levels at 24 mills and would constrain to some extent 1994 emission increases at another nine mills. It would reduce aggregate 1994 TSP emissions from pulp-producing mills from 174,000 tons to 158,000 tons -- a 15 percent decrease. The 33 mills could meet this emission reduction at an annual cost of approximately $19.4 million, which would increase 1994 annual TSP pollution control costs by 15 percent.

In order to assess whether a nondegradation approach would result in net benefits, we started by examining its air quality implications. The general implications of such an approach can be simulated with the MCP air quality model. The application of the MCP model indicates how the increase in emissions from the 33 mills, in the absence of a nondegradation approach, would affect ambient air quality and how ambient air quality would improve with the nondegradation approach.

Actually, the MCP model could estimate potential air quality concentrations resulting from TSP emissions for 27 of the 33 mills that expand capacity and that would be subject to more stringent emission limits (Figure 7-4). The model could not simulate air quality concentrations for the other six mills because meteorological data have not yet been incorporated into the MCP model for all areas of the country.

Figure 7-4. Air Quality Impacts and Pollution Reduction Costs for the 54 Mills Potentially Affected by a 1994 Nondegradation Approach for TSP (Annual Costs -- $1984 10^6)

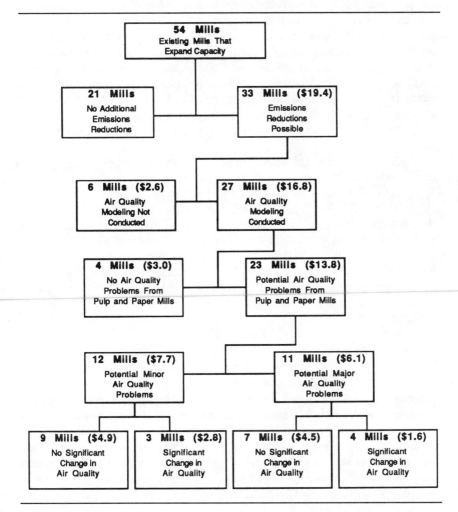

The MCP modeling indicates that 1994 emissions from 23 mills could contribute to air quality problems because the ambient concentrations around the mills would exceed 40 $\mu g/m^3$. The other four mills are judged not to contribute to air quality problems because the ambient concentrations around the mills would not exceed 40 $\mu g/m^3$.

Adopting a nondegradation approach at the 23 mills that are potentially contributing to air quality problems would reduce TSP emissions, but would not necessarily result in a "significant" improvement in air quality. To refine the analysis, we followed the same procedure and criteria used to evaluate a technology-forcing approach. We first separated the potential air quality problems into minor and major problems. As a result, 12 mills are categorized as contributing to minor air quality problems, and 11 mills are categorized as contributing to major air quality problems. We next separated air quality changes resulting from imposition of this approach into significant and non-significant improvements. The MCP modeling indicates that emission reductions at seven mills would result in significant emission reductions. Of these seven mills, four mills are emitting TSP that could cause major air quality problems and three mills are emitting TSP that could cause minor air quality problems.

We next applied the benefit-cost framework developed in Chapter 4 to the seven mills where emission reductions would result in significant air quality changes (Table 7-7). In the aggregate, imposition of this approach at the seven mills would not result in a positive benefit-cost comparison. The monetary value of the reductions in diverse health and welfare effects range from $0.3 million to $1.7 million, and the costs to achieve this result are $4.5 million. A closer examination of the findings reveals that the maximum estimate of monetary benefits of air quality improvements exceed the costs of pollution reduction at only one of the seven mills. At the other six mills, the costs of emission offsets exceed the benefits.

An obvious conclusion emerges from this evaluation of a nondegradation approach. Extrapolating from our evaluation of emission reductions at 27 mills, this approach would not result in a reasonable balance between health and welfare benefits and pollution reduction costs. The emission reductions from this approach would significantly improve air quality around only nine of the 33 mills; reductions would not improve air quality around the other 24 mills. Economic benefits would occur at only nine mills, and the benefits would range from $0.4 million to $2.1 million. The costs of this approach -- $19.4 million -- would exceed the discernible air quality improvements and measurable economic benefits.

Table 7-7. Benefit-Cost Analysis for the Seven Mills Affected by a
1994 Nondegradation Approach -- TSP ($1984 10^3)

Mill	Air Quality	Tons Reduced	Popula-tion	Annual Benefits Min	Mid	Max	Annual Costs[a]	Benefits >Costs?
713	major/s[b]	670	22,000	$10	$30	$80	$520	No
807	major/s	330	36,000	10	20	50	570	No
847	major/s	690	9,500	0	0	10	50	No
852	major/s	340	10,000	0	0	10	550	No
708	minor/s	390	10,500	0	10	20	660	No
733	minor/s	410	513,600	240	670	1,460	780	Yes
797	minor/s	780	11,100	0	10	20	1,350	No
TOTAL		3,610	612,800	$260	$740	$1,650	$4,480	

a. Based on estimated 1994 capacity.
b. s = Significant improvement in air quality (0.1 $\mu g/m^3$ or greater).

Benefits Approach

Just as with the above analyses of different approaches for water
pollution management, a consistent result emerges from the efficiency
evaluations of both technology-forcing and nondegradation approaches for
future air pollution management. The uniform implementation of both
approaches would not produce net benefits, but implementation of these
policies at a subset of the mills would. This result suggests that we turn
to the third generic approach, a benefits approach, as the preferable
option for mitigating the adverse environmental effects of criteria air
pollution in the future. A benefits approach would result, by design, in
net benefits. If consistently pursued, such an approach would result in a
reasonable balance between the benefits and costs of eliminating violations
of ambient air quality standards.

A benefits approach builds upon ambient air quality conditions. It
starts with air quality conditions and the physical potential for
improvements in air quality as a result of emission reductions. Its unique
feature is that it incorporates the potential for reducing human exposure
as well as air quality improvements in designating which mills should

reduce their emissions. As such, it limits concern for ambient improvements only to those mills where the benefits of pollution reduction would exceed the control costs.

Our evaluation of the efficiency consequences of a benefits approach began with an inventory of the pulp and paper mills potentially causing air quality problems. We compiled the inventory from our analysis of technology-forcing and nondegradation approaches. We took eight mills that are subject to the technology-forcing approach but that would not increase their capacity between 1984 and 1994 from Table 7-6, and we took seven mills subject to the nondegradation approach and that would increase their capacity from Table 7-7 to form a total inventory of 15 mills.[6] These mills potentially cause minor and major air quality problems, and reducing emissions from these mills would significantly, as defined earlier, improve air quality.

We next applied the benefit-cost framework to the 15 mills where reduction of emissions would result in a significant improvement in air quality (Table 7-8). We eliminated 10 mills because the control costs would exceed the anticipated benefits, primarily reductions in adverse health effects. Thus, only five mills would be targeted for emission reductions under a benefits approach. With annual costs of $3.6 million and potential benefits of $2.0 million to $11.4 million, the benefit-cost ratio of this approach would be almost two.

Findings -- TSP

In summary, the conventional approaches that EPA has traditionally pursued to minimize the environmental risks of pollution resulting from capacity expansion in the pulp and paper industry would not result in a reasonable balance between health and welfare benefits and pollution reduction costs (Table 7-9). Moreover, a technology-forcing approach, the less wasteful in a benefit-cost sense, would constrain effluent increases at only one-half of the mills that increase their capacity.

The preferable approach for reducing environmental risks is a benefits approach. This approach would require only a subset of the mills subject to either a technology-forcing or a nondegradation approach to reduce their emissions. The subset of mills would include only those individual mills where the benefits exceed the costs. Our analysis of that subset shows a slightly positive benefit-cost ratio. The policy is more successful than the other two policies because it starts with environmental problems

Table 7-8. Benefit-Cost Analysis for the 15 Mills Potentially Affected by
a 1994 Benefits Approach -- TSP ($1984 10^3)

Mill	Air Quality	Tons Reduced	Population	Annual Benefits Min	Annual Benefits Mid	Annual Benefits Max	Annual Costs[a]	Benefits >Costs?
No Increase								
728	major/s[b]	2,800	164,800	$1,400	$3,400	$7,700	$1,220	Yes
736	major/s	1,440	6,500	180	510	1,120	660	Yes
776	major/s	660	9,700	0	0	10	450	No
791	major/s	920	12,700	170	480	1,050	1,400	No
811	major/s	1,500	15,400	50	150	330	1,900	No
862	major/s	120	40,100	80	210	450	320	Yes
724	minor/s	330	79,800	90	260	580	570	Yes
751	minor/s	2,000	132,000	50	150	330	450	No
Increase								
713	major/s	670	22,000	$10	$30	$80	$520	No
807	major/s	330	36,000	10	20	50	570	No
847	major/s	690	9,500	10	20	40	550	No
852	major/s	340	10,000	0	0	10	550	No
708	minor/s	390	10,500	0	10	20	660	No
733	minor/s	400	513,600	240	670	1,560	780	Yes
797	minor/s	780	11,100	10	30	70	1,350	No
Subtotal (5 mills with B>C)		5,090	804,800	$1,990	$5,050	$11,410	$3,550	Yes
Total (15 mills)		13,370	1,073,800	$2,300	$5,940	$13,400	$11,950	No

a. Based on estimated 1994 capacity.
b. s = Significant improvement in air quality (0.1 ug/m^3 or greater).

and targets additional pollution reduction to areas where there is potential
for significant improvement in air quality and a sufficiently large
population to benefit from the reductions in health risks. The Clean Air
Act with its emphasis on meeting ambient air quality standards per se does
not encourage either EPA or the states to pursue such a policy, however.

DESIGNATED AIR POLLUTANTS -- TRS

In this section, we cursorily estimate the benefits and costs of three
approaches to reduce TRS emissions during the 1984-1994 period covered
by this study. As described at the beginning of the chapter, TRS

Table 7-9. Summary Comparison of 1984 - 1994 Regulatory Approaches for TSP ($1984 10^6)

Approach	Number of Mills Affected	Annual Benefits[a]	Annual Costs[b]	Benefit-Cost Ratio
Technology-Forcing	59	$3.0 - $17.0	$49.1	0.2
Nondegradation	33	$0.4 - $2.1	$19.4	<0.1
Benefits	5	$2.0 - $11.4	$3.6	1.4

a. Benefit estimates are extrapolated to reflect all the mills that would be affected by the regulatory approach.
b. Based on estimated 1994 capacity.

emissions are estimated to increase by approximately 1,000 tons in spite of the requirement to meet New Source Performance Standards emission limits at expanding and new mills. The potential regulatory approaches -- technology, ambient, and benefits -- are those that we have used to define regulatory approaches for other pollutants.

This evaluation of TRS emission reduction approaches is more tentative than the evaluation of TSP emission reduction approaches. We do not have data for preparing mill-specific cost estimates for emission reductions more stringent than those required to meet New Source Performance Standards. Instead, we used a generalized engineering estimate of the costs of installing ductwork to capture TRS emissions and to route them through the black liquid oxidation furnace. Such ductwork would require a capital investment of between $0.5 to $5.0 million, and it would offset most, if not all, total TRS emission increases.[7] This evaluation is also more limited than the one for TSP. We focused only on existing mills whose capacity expansion, in spite of meeting New Source Performance Standards emission limits, would increase their TRS emissions by more than five tons per year. As a result, we limited our analysis to 32 of the 42 kraft mills with expanded capacity to produce kraft pulp (Appendix 7-A).

Technology-Forcing Approach

The 32 mills with additional capacity to produce kraft pulp would increase TRS emissions by 600 tons per year even after meeting New Source Performance Standards. On average, TRS emissions at the 32 mills would be approximately 30 percent greater than in 1984. At six of the mills, however, TRS emissions would be 50 percent greater and at four mills, they would be 100 percent greater. As can be seen in Table 7-10, we estimate that these total emission increases could be eliminated at an annual cost between $2.1 and $20.8 million, with a mid-point estimate of $9.3 million.

The benefits of constraining TRS emission increases are estimated just as we estimated the benefits of TRS emission reduction as a result of the Clean Air Act. We calculated the number of days that households would be exposed for at least one hour to noticeable TRS (above 0.005 mg/m^3) under the expected growth conditions. We then calculated the percentage decrease in such days if total TRS emission were constrained to their 1984 rate. On the basis of this percentage decrease, we estimate that the benefits of constraining TRS emissions to their 1984 level would range from $3.6 to $22.2 million with a mid-point estimate of $9.3 million.

In the aggregate, implementing a technology-forcing approach at the 32 mills would generate net benefits. The monetary value of reductions in days exceeding the odor threshold would range from $3.6 to $22.2 million, and the costs of emission reductions to achieve this result would range from $2.1 to $20.8 million. The mid-point benefit and cost estimates are the same, $9.3 million. A closer examination of the findings, however, reveals that the monetary benefits of TRS emission reduction exceed the costs at approximately 11 out of the 32 mills. At the other 21 mills, the costs of a technology-forcing approach would exceed the benefits.

Ambient Approach

The Clean Air Act does not require EPA to set an ambient standard for TRS.[8] As indicated in Chapter 5, however, there is general consensus that TRS odors above 0.005 mg/m^3 are detectable. In order to simulate the effects of an ambient-based policy, we arbitrarily set a one-hour ambient standard at 0.05 mg/m^3, an order of magnitude greater than the odor threshold. We then calculated the highest and second highest TRS concentrations resulting from the 1994 TRS emissions. We used

Table 7-10. Benefit-Cost Analysis for Selected Mills with Increased
TRS Emissions Between 1984 and 1994 ($1984 10^6)

Approach	Number of Mills Affected	Annual Benefits	Annual Costs	Mills Where Benefits > Costs	Benefit-Cost Ratio
Technology-Forcing	32	$9.3 (3.6-22.2)	$9.3 (2.1-20.8)	11 (6-19)	1.0
Ambient	25	$7.4 (2.5-20.3)	$7.3 (1.6-16.2)	9 (5-18)	1.0
Benefits	11	$8.1 (5.5-9.1)	$3.2 (1.2-3.2)	11 (5-19)	2.5

the MPTER model, just as we did in Chapter 5, to simulate TRS concentrations. We found that 25 of the 32 mills would exceed the assumed ambient standard, using the second highest TRS concentrations as an indicator.

As can be seen in Table 7-10, we estimated that the incremental emission increases between 1984 and 1994 could be eliminated with additional ductwork at the 25 mills. The annual cost would be between $1.6 and $16.2 million. Just as we did for the technology-forcing approach, we also estimated the benefits of constraining TRS emissions to their 1984 level at the 25 mills. The monetary benefits would range from $2.5 to $20.3 million.

In the aggregate, implementing an ambient approach would generate net benefits. The monetary benefits of 25 mills constraining their potential TRS emission increases would range from $2.5 to $20.3 million, and the costs of emission reductions to achieve this result would range from $1.6 to $16.2 million. The mid-point benefit and cost estimates are essentially the same. In fact, the results from the technology-forcing and the ambient approaches are essentially the same. Similarly, a closer examination of the finding reveals that the monetary benefits of TRS emission reductions would exceed costs at approximately one-third of the mills. At the other 16 mills, the costs of complying with a suggested ambient approach would exceed the benefits.

Benefits Approach

Although both the technology-forcing and ambient approaches would generate in the aggregate net benefits, they would require pollutant reductions at many mills where the costs would exceed the benefits. As we have learned from our examination of alternative approaches for reducing other pollutants, a more efficient approach would be a benefits approach. The approach would be based on mill-specific evaluations of benefits and costs and would require pollutant reductions at only those mills where benefits would exceed costs.

In our analysis of the technology-forcing approach, we calculated the benefits and costs at each mill. An examination of the data revealed that TRS emission reductions at approximately 11 mills would result in benefits exceeding costs. (The actual number could be as few as five and as many as 19.) The benefits would be approximately $8.1 million, and the costs would be approximately $3.2 million.

Findings -- TRS

Our findings about potential regulatory approaches for reducing increased TRS emissions are decidedly different from our findings about regulation of other pollutants. Only in the case of TRS emission reduction would all three potential regulatory approaches generate net benefits. (For water and TSP air, only a benefits approach would generate net benefits.) A benefits approach for TRS, however, would be preferable to the other two policies because benefits are approximately twice the costs compared to the other two policies, where the benefits are approximately equal to the costs.

In summary, an aggressive regulatory program to constrain the growth in TRS emission during the period of 1984 to 1994 would appear to be warranted. Any of three potential regulatory approaches would produce, in the aggregate, reductions that would result in net benefits. In addition, a benefits approach for TRS compares much more favorably on efficiency grounds than the one for TSP.

FINDINGS

The findings about the three regulatory approaches for reducing the environmental risks associated with industrial growth generally make a strong argument for a benefits approach. The benefits approach is the only approach that would result in significant net benefits for both conventional water and criteria air pollutants. Neither the technology-forcing nor antidegradation approach would generate net benefits. Additionally, the benefits approach is the most efficient approach for reducing the environmental risks associated with increased levels of TRS emissions.

The positive findings about all three approaches for TRS emissions suggests that regulating TRS should receive priority among policies for reducing environmental risks from currently regulated pollutant releases from pulp and paper mills. Apparently, increased TRS emissions from all mills that expand will be a considerable nuisance to the affected populations, and the costs of emissions constraints, even if based on a uniform technology approach, would generate net benefits.

The findings about the ambient approach deserve special attention because they contradict the findings about the efficiency of the ambient approach for the 1973-1984 period. We speculate that the ambient approach does not appear to be as efficient in future projections as it has been in past practice because the analysis does not take into account the behavior of local environmental managers. Our impression from discussions with those managers dealing with air pollution problems is that they take into account variations in cost-effectiveness when using EPA's Control Technology Guidance to set source specific emission limitations.

Notes

1. We projected the estimated capacity increase in 1986. More current information suggests that pulp capacity will increase by 12,000 rather than 8,100 tons and that paper and paperboard capacity will increase by 17,000 rather than 9,800 tons [API. 1989]. This substantial increase only reinforces the importance of EPA developing an explicit policy for mitigating the environmental consequences of industrial growth.

2. The only difference in the benefit-cost evaluation in this chapter compared to Chapter 3 is that we ignored as a "significant" change in water quality any within use category change that affects less than 25 percent of the stream mileage in that use category.

3. For the benefit analysis, we estimated the 1994 population by assuming that population would increase by 0.8 percent annually between 1985 and 1994 based on Bureau of the Census projections [U.S. Bureau of the Census. 1984].

4. We explored the possibility of evaluating the efficiency consequences of potential policies for reducing environmental risks from SO_2 emissions, but decided that acid rain legislation would have only minimal impact on the pulp and paper industry. First, SO_2 emissions from coal-fired boilers at pulp and paper mills accounted for only about two percent of all SO_2 emissions in the United States in 1985 and about 15 percent of SO_2 emissions from industrial (non-utility) sources [EPA. 1988]. Second, the costs per ton of SO_2 reduced appeared to be very high relative to the costs of SO_2 emission reductions at utility and other industrial sources. EPA analysis of acid rain legislation in the 100th Session of Congress reported the cost per ton of SO_2 reduced for a 10- to 12-million-ton reduction to be approximately $500 [ICF, Inc. 1988]. Based on the work of E.H. Pechan and Associates [1986], we estimated the costs of a 200,000 ton-reduction for coal-fired boilers at pulp-producing mills with flue gas desulfurization. The per ton cost of SO_2 emissions for 105 coal fired boilers, assuming 90 percent scrubber efficiency, would be $2,770. Only nine boilers could reduce SO_2 emissions for less than $1,000 per ton, and only one boiler (at a mill with three boilers) could reduce SO_2 emissions for less than $500 per ton. Based on these findings, we decided that evaluation of alternative policies for regulating the pulp and paper industry would contribute very little to the national debate about appropriate policies for future SO_2 emission reductions.

5. The median change at the 49 mills with a decrement greater than 1.0 $\mu g/m^3$ was 2.0 $\mu g/m^3$.

6. The criterion for inclusion in the inventory is more expansive than for the water analysis because we had fewer mills in our initial inventories due to data limitations.

7. This generic cost estimate was provided by Neil McCubbin, a consulting engineer to the pulp and paper industry. [McCubbin. 1989].

8. The legislative history for the 1970 Amendments to the Clean Air Act reveals that Congress considered listing TRS as a criteria air pollutant, an action what would have required EPA to set an ambient standard.

References

American Paper Institute (API). 1989. *Paper Paperboard Woodpulp Capacity.* New York, NY.

Baum, Jonathan K. 1983. "Legislating Cost-Benefit Analysis: The Federal Water Pollution Control Act Experience." *Columbia Journal of Environmental Law*, 9:75-111.

Code of Federal Regulations (CFR). 1985. "General Provisions; Definitions." 51:1(o), Appendix B. U.S. Government Printing Office, Washington, DC.

Grad, Frank P. 1985. *Treatise on Environmental Law*, 1:2-160. Matthew Bender and Co., New York, NY.

E.H. Pechan and Associates. 1986. "Acid Deposition Control Techniques for the New England States," Report to Northeast States for Coordinated Air Use Management, Boston, MA. Springfield, VA.

Federal Register (FR). 1978. "Standards of Performance for Stationary Sources; Kraft Pulp Mills." 43:7568-7596.

_____. 1986(a). "Pulp, Paper and Paperboard and the Builders' Paper and Board Mills Point Source Categories; Best Conventional Pollutant Effluent Limitations Guidelines." 51:45232-45242.

_____. 1986(b). "Standards of Performance for New Stationary Sources; Industrial-Commercial-Institutional Steam Generating Units." 51:42768-42796.

ICF Inc. 1988. "Preliminary Analysis of the Mitchell Bill (S. 1894)," Report to the U.S. Environmental Protection Agency. Fairfax, VA.

McCubbin, Neil. 1989. Personal communication with the author. Fishers Point, Foster, Quebec, Canada.

Putten, Mark C., and Jackson, Bradley D. 1986. "The Dilution of the Clean Water Act." *Journal of Law Reform*, 19:863-901.

United States Code, 42, Section 7401 (b) (1). 1982 Edition.

United States Code, 33, Section 101. 1982 Edition.

U.S. Bureau of the Census. 1984. "Projections of the Population of the United States, by Age, Sex and Race: 1983 to 2080." Current Population Reports, Series P-25, Number 952. U.S. Government Printing Office, Washington, DC.

U.S. Environmental Protection Agency. 1985. "National Emissions Data System." Office of Air Quality Planning and Standards, Research Triangle Park, NC.

_____. 1986. "Development Document for Best Conventional Pollutant Control Technology Effluent Limitations Guidelines for Pulp, Paper and Paperboard and Builders' Paper and Board Mills Point Source Categories." EPA-400/1-86-025. Washington, DC.

_____. 1988. "Anthropogenic Emissions Data for 1985 NAPAP Inventory." EPA-600/7-88-022. Research Triangle Park, NC.

Appendix 7-A

1984-1994 TRS EMISSION REDUCTION POLICIES

Table 7-A-1. Mill Specific TRS Benefit and Cost Estimates for 1984-1994 ($1984 10^6)

Mill	Annual Benefits			Annual Costs		
	Min	Mid	Max	Min	Mid	Max
700	$ 0	$ 0	$ 560	$ 70	$ 290	$ 650
702	0	30	50	70	290	650
703	100	160	180	70	290	650
704	280	560	840	70	290	650
708	170	260	260	70	290	650
713	0	0	180	70	290	650
720	0	20	150	70	290	650
722	50	70	70	70	290	650
723	200	340	360	70	290	650
725	0	420	830	70	290	650
730	0	180	670	70	290	650
731	210	320	320	70	290	650
733	0	0	4,280	70	290	650
735	0	800	1,840	70	290	650
737	0	0	200	70	290	650
743	0	780	2,260	70	290	650
744	1,050	2,190	3,350	70	290	650
748	80	130	150	70	290	650
759	0	0	170	70	290	650
761	290	600	920	70	290	650
783	0	0	260	70	290	650
797	0	0	90	70	290	650
798	0	30	90	70	290	650
807	500	800	850	70	290	650
808	0	0	290	70	290	650
810	0	140	300	70	290	650
816	60	95	120	70	290	650
837	0	350	710	70	290	650
848	0	0	440	70	290	650
850	0	140	350	70	290	650
854	0	0	70	70	290	650
888	610	930	960	70	290	650
TOTALS						
32 Mills	$3,600	$9,300	$22,200	$2,100	$9,300	$20,800
25 Mills	$2,500	$7,400	$20,300	$1,600	$7,300	$16,200
11 Mills	$5,500	$8,100	$9,100	$1,200	$3,200	$3,200

Chapter 8

TOXIC POLLUTANTS

Under the Clean Water Act, Clean Air Act, and Resource Conservation and Recovery Act, EPA has the authority to regulate toxic pollutants emitted by industrial sources. The Clean Water Act of 1977 requires EPA to set standards for a specific list of 65 toxic pollutants and classes of pollutants that potentially includes thousands of specific pollutants. The Clean Air Act of 1970 requires EPA to set standards for hazardous air pollutants that may cause adverse health effects. The Resource Conservation and Recovery Act of 1984 requires EPA to identify hazardous wastes and to set standards for facilities that generate or manage hazardous wastes. As of 1989, EPA had established toxic water pollutant limitations for three chemicals discharged by the pulp and paper industry, but no hazardous air or hazardous waste pollutant limitations.

Among the unregulated toxic pollutants emitted from pulp and paper mills are several total organochlorines (TOCLs).[1] This chapter reviews potential regulations for two of the TOCLs, dioxin and chloroform. It discusses dioxin primarily as a water pollution problem, and chloroform primarily as an air pollution problem and secondarily as a hazardous waste constituent.

To the extent permitted by available data, this chapter continues our critique of technology, ambient, and benefits approaches for reducing environmental risks. The toxic pollutant provisions in the Clean Water Act (Section 307) combine technology and ambient (health) considerations; EPA interprets the statute as an ambient approach. The hazardous air pollutant provisions in the Clean Air Act (Section 112) focus on technology considerations; EPA interprets the statute as a technology approach tempered by de minimis ambient (health) risks. The hazardous waste provisions in the Resource Conservation and Recovery Act (Section 3001) focus on technology considerations; again, EPA interprets the statute as a technology approach tempered by de minimis ambient (health) risks. None of the provisions nor their interpretation by EPA appears to allow regulations that would be based upon net benefits. Here, we offer some regulatory options based on a benefits approach and contrast them with technology and ambient approaches.

WATER POLLUTANTS -- DIOXIN

EPA has known for several years that the widespread use of chlorine and its derivatives for bleaching kraft and sulfite pulps causes discharges of chlorinated organic compounds. This family of substances is frequently referred to as organochlorines (chlorine bound with organic compounds). Included among these organochlorines are small amounts of the more toxic chlorinated dioxins and the less toxic chlorinated furans. These two families of chemicals, the chlorinated dibenzo-p-dioxins and the chlorinated dibenzo-furans, include 210 substances -- 75 different dioxins and 135 furans. Two of these substances, 2,3,7,8-tetrachloro-p-dibenzodioxin (2378-TCDD) and 2,3,7,8-tetrachlorodibenzofuran (2378-TCDF), are particularly important from a toxicology point of view. The substance 2378-TCDD is the most toxic of these substances and is the one generally meant in references to "dioxin" in the press. Unless otherwise indicated, we will use the term "dioxin" only if it is 2378-TCDD.

The organochlorine content of pulp industry wastewaters is generally thought to be best measured by the American Public Health Association Test Method 506 [APHA. 1985]. Strictly, this method measures total organic halides (TOX). Because chlorine is the only halide present in kraft and sulfite wastes in significant quantities, TOX for all practical purposes is identical to total organochlorines. Less than one ten-millionth of the organochlorines in kraft and sulfite waste is made up of 2378-TCDD. The concentration of 2378-TCDD is measured in parts per trillion (ppt).

Effluents

There has been a significant increase in the production of bleached pulp and in industry consumption of chlorine since the 1950s. Total wood pulp capacity increased from 40.3 million tons in 1966 to 59.6 million tons in 1985, and to 64.3 million tons for 1989 [API. 1968, 1985, 1989]. Of these totals, 35 percent consisted of bleached pulp in 1966, 42 percent in 1985, and an estimated 45 percent in 1989.

Very few data have been published on the organochlorine discharges from pulp mills. Most measurements have been performed since early

1988. Bonsor, McCubbin, and Sprague [1988] in their Ontario study used a modification of Germgard's equation relating organochlorine discharges to chemical consumption [Germgard and Larson. 1983].[2]

Using a modification of Germgard's equation, McCubbin [1988] calculated the per ton and total discharge of TOX for 103 of the 104 mills in the EPA/API study. As illustrated in Figure 8-1, there is considerable variability among mills. The average discharge is approximately 5.4 pounds of TOX per air-dried ton of pulp, and the total discharge from the 103 mills is 193 tons per day.

Figure 8-1. TOX Discharges for 103 Bleached Kraft Mills

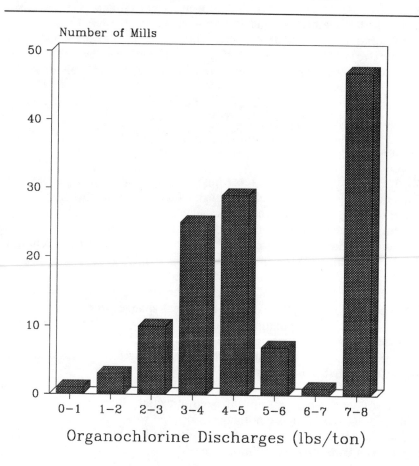

Health and Ecological Effects

Experimental studies with 2378-TCDD in animal systems have demonstrated a variety of toxic effects [EPA. 1985(a)]. These effects include carcinogenesis, cancer promotion, reproductive and teratogenic effects, immunotoxic effects, thymus atrophy, liver damage, and effects on the skin and thyroid. Limited toxicological testing of other chlorinated dioxins and chlorinated furans has demonstrated that several of these compounds cause similar toxicological effects, but that higher doses of these compounds are generally required to cause effects of comparable magnitude to those induced by 2378-TCDD.

The presence of 2378-TCDD in aquatic and terrestrial organisms has been associated with lethal, carcinogenic, teratogenic, reproductive, mutagenic, histopathologic, and immunotoxic effects [Eisler. 1986]. There are substantial inter- and intra-species differences in sensitivity and toxic responses to 2378-TCDD. Typically, animals poisoned by 2378-TCDD exhibit weight loss, atrophy of the thymus gland, and eventually death.

One of the most extensive surveys of available data on aquatic toxicity of pulp and paper mill effluents was done in response to the dioxin issue by McLeay and Associates Limited [1987]. Following are a few of the conclusions of particular relevance to this issue.

1. The vast majority of receiving-water studies as of 1986 did not distinguish biological effects due to toxic effluent constituents from those caused by other effluent characteristics such as temperature, color, salinity, nutrients, solids concentration, and oxygen demand. Such a distinction is necessary for the rational determination of the required degree of removal of toxic constituents from an effluent.

2. The preponderance of receiving-water studies associated with pulp and paper mill discharges examined biological changes in communities of benthic invertebrates, photoplankton, and periphyton. Although many site-specific instances of effects on these communities have been found, few if any of these studies have shown that these changes resulted in adverse effects on the fisheries resource.

3. Instances where significant environmental effects on aquatic life have been demonstrated are normally associated with pulp and paper mills discharging untreated or primary-treated effluent into poorly flushed waters with limited capacities for dilution and dispersal.

4. No studies were found that showed conclusive evidence for significant toxic effects within receiving waters that were attributable to the discharge of typical biotreated mill effluent.

Over the past 10 years, the Swedish government and industry have studied the ecological effects of bleached pulp wastes [Sodergren et al. 1987]. They carried out aquatic studies in the vicinity of two mills, one producing unbleached kraft and the other bleached kraft. They documented deleterious effects on the biota at distances up to six miles from the bleached kraft mill. They reported erosion of the fins of fish and failure or delay of gonad development.

Ambient Concentrations

We modeled the water quality impacts of dioxin discharge from the 74 mills that discharge into rivers [Grayman. 1989]. We assumed a range of discharge concentrations in the effluent based on the "5-mill" study [EPA. 1988(d)], which reported a range of 0.015 to 0.12 ppt. We took 10 percent of 0.015 ppt as the low value and 0.12 ppt as the high value.[3] Figure 8-2 shows the percentage of mills where the resulting in-stream concentrations exceed EPA's water quality criterion for human health protection under different stream flow and concentration conditions [EPA. 1984]. As illustrated, the analysis indicates a significant problem even at the low end of the range (with average stream flow and 10 percent of the reported low concentration). The figure also shows the percentage of mills where the resulting in-stream concentrations exceed the Fish and Wildlife Service's recommended water quality criterion for aquatic life protection [Eisler. 1986]. Even though the Fish and Wildlife criterion is almost 1,000 times greater than the human health criterion, there is still a substantial number of exceedances, especially at low flow and/or with the higher range of loading assumptions.

Risks to Public Health

The evidence for the carcinogenic (cancer-causing) action of 2378-TCDD is provided mainly by several long-term studies of laboratory animals exposed to the substance. On the basis of these animal studies and associated factors, EPA concluded that 2378-TCDD is an animal carcinogen and should be regarded as a probable human carcinogen [EPA. 1985(a)]. Applying its established procedures, EPA used the experimental animal data to estimate an upper bound on the cancer potency factor

Figure 8-2. Percentages of Mills Violating In-Stream Dioxin Standards

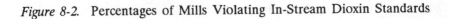

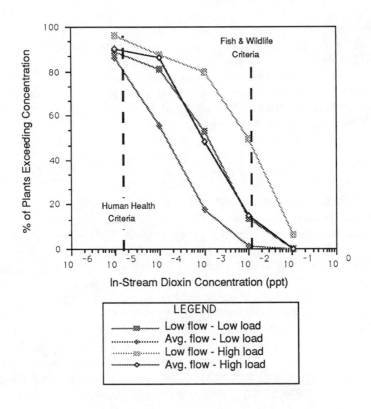

In-Stream Dioxin Concentration (ppt)

LEGEND
Low flow - Low load
Avg. flow - Low load
Low flow - High load
Avg. flow - High load

a. Analysis based on dioxin loads calculated as a function of discharge concentrations.

(referred to as "q1*") for 2378-TCDD of 1.6 x 10^{-5} per mg/kg/day. Although this number remains EPA's current reference value on the potency of 2378-TCDD, EPA recently completed a formal reassessment of the cancer potency factors [EPA. 1988(a)]. Preliminary results suggest that the current factor overestimates the cancer risk.

2378-TCDD has been shown to be teratogenic (causing fetal malformations) and to cause adverse reproductive effects in a number of animal species, including subhuman primates [EPA. 1985(a)]. EPA has evaluated the potential for non-cancer effects from long-term exposure to dioxins and has calculated an estimate of one pg/kg/day as the daily exposure that is likely to be without risk of adverse health effects. We do not incorporate this EPA estimate of non-cancer health in our risk assessment.

We could not review all the pathways of human and ecosystem exposure to dioxin from pulp-producing mills. These pathways include vapor and dust inhalation by people working and living around the mills. Instead, we focused on what appears to be a major pathway -- dioxin-contaminated sediments in the aquatic environment. We excluded another potentially major pathway -- dioxin-contaminated paper. Although EPA has examined the exposures and risks that may occur through use of bleached kraft paper products, the results are very uncertain, given the underlying assumptions [Arthur D. Little. 1987]. The results of the ongoing, court-mandated EPA risk assessment may resolve some of the uncertainties in EPA's preliminary assessment [FR. 1985].

Our approach for characterizing risks focused on human consumption of contaminated fish and followed the EPA risk assessment for dioxin contamination at the Dow Chemical Plant in Midland, Michigan [EPA. 1988(c)]. Our approach used the EPA cancer potency factor, the dioxin residue levels found in fish downstream from 50 of the 104 mills [EPA. 1989], and fish consumption patterns identified by EPA's Exposure Assessment Group [EPA. 1988(b)]. Table 8-1 presents the upper-bound estimates of excess lifetime cancer risks. These cancer risk estimates range from one chance in 1,000 for the reasonable worst case with the highest reported level of dioxin to three chances in 1 million for the general consumer with the minimum reported level.

We estimated the population potentially consuming fish from these streams just as we estimated the potential recreation population -- i.e., all population within a 30-mile radius of the site times a visitation rate based on the stream mileage in that area. For the 74 modeled sites, the population potentially catching and consuming fish from these streams is approximately 15 million. We think that the median and minimum fish concentration and the general consumption pattern, which corresponds to about a quarter-pound of contaminated fish every two weeks, is an upper-bound estimate. The maximum individual risk of contracting cancer in a lifetime (assumed to be 70 years) would be between 3.0 chances in 100,000 and 3.0 chances in 1,000,000. The resulting annual population cancer incidence would be between 0.6 and 6.0.

Table 8-1. Risk Assessment for Ingestion of Fish Contaminated
with 2,3,7,8-TCDD

Fish Concentration[a] (pg/g)	Fish Consumed/Day[b] (g/day)	Upper Bound Daily Dose Rate[c] (pg/kg/day)	Cancer Risk[d]
16.08	6.5	1.49	$2.0E^{-4}$
	30.0	6.89	$1.0E^{-3}$
1.75	6.5	0.16	$3.0E^{-5}$
	30.0	0.75	$1.0E^{-4}$
0.17	6.5	0.02	$3.0E^{-6}$
	30.0	0.07	$1.0E^{-5}$

a. The fish concentration levels are the maximum, median, and minimum levels
reported in EPA's ongoing National Bioaccumulation Survey (March 1989).
b. Fish consumption of 6.5 g/day represents the general consumer, and 30g/day
represents an individual who consumes larger quantities than the national
average based on recreational fishing [EPA, 1988(b)].
c. Daily Dose Rate = Fish Concentration x Fish Consumed per Day/80kg.
d. Upper Bound Cancer Risk = Daily Dose Rate (pg/kg/day) x Cancer Potency.

Risks to Ecosystem

EPA has not published water quality criteria for protecting aquatic
life. According to the Fish and Wildlife Service, however, the limited data
suggest that dioxin concentrations in water should not exceed 0.01 ppt to
protect aquatic life or 10 to 12 ppt in food items of birds and other
wildlife [Eisler. 1986].

Returning to our modeling of dioxin concentrations downstream from
74 mills, we looked for situations where dioxin concentrations exceeded the
aquatic life standard. Whereas the water quality modeling showed most
of the mills exceeding the human health standard under a range of
discharge scenarios, it showed only 50 percent of mill discharges exceeding
the aquatic life standard with low flows and high concentrations and only
20 percent exceeding this standard with low flows and low concentrations
in the discharge. Given the lack of information about aquatic ecosystems
downstream from bleached kraft mills and the uncertainty and variability
in adverse effects, we could not characterize the aquatic ecosystem risks.

According to the Fish and Wildlife Service, growth reduction in northern pike and rainbow trout was observed at 0.01 ppt [Eisler. 1986].

To determine if fish-eating birds are jeopardized by dioxin from the 104 mills, we used the upper-bound threshold of 12 ppt and the dioxin concentration in fish downstream from 50 mills [EPA. 1989(b)]. We used the whole-fish analysis rather than fillets analyses because fish-eating birds often consume the entire fish.   Carp is the species most likely to be hazardous to fish-eating birds. Dioxin concentrations in carp exceeded the standard downstream of six mills in six states.   Dioxin concentrations in catfish, pike, and suckers exceeded the standard downstream of five mills in five states.   Given the geographic diversity of sites with excessive concentrations in fish, we think that there is widespread potential for adverse ecological effects.

Effluent Reduction Technology and Costs

The widespread use of chlorine and its derivatives for bleaching kraft and sulfite pulps generates chlorinated organic compounds. Pulps mills reduced their discharge by approximately 30 percent when they installed biological treatment to comply with the Best Practicable Technology standards. To achieve further reductions, pulp mills must shift away from the use of chlorine in the bleaching process. (See Appendix A for a description of the bleaching process.)

One approach for reducing the discharge of chlorinated organic compounds is substitution of chlorine dioxide for chlorine in the first bleaching stage [Bonsor, McCubbin, and Sprague. 1988]. At present, the industry believes that some chlorine is needed to make fully bleached pulp, but the amount of chlorine could be as little as 25 percent rather than the current practice of 85 to 100 percent. This approach is well established and already used by about 20 mills in the United States as well as many others worldwide. Another approach would lower the need for chlorine by removing more lignin in other ways. A potential option for extended delignification is oxygen bleaching, which involves intensive mixing of oxygen with the pulp before the bleaching sequence. Clearly, a third approach is to reduce the degree of brightness, which would reduce the degree of bleaching. Yet, in the main, industry has gone in the opposite direction from pollution prevention with such products as white corrugated cartons.

EPA has not yet prepared a report on the cost and effectiveness of alternative process changes that would reduce the discharge of chlorinated organic compounds. Consequently, we turned to one of the authors of a recent Canadian report [Bonsor, McCubbin, and Sprague. 1988] and public information about the kraft and sulfite bleach lines in the 104 mills to prepare our own assessment. Instead of following the usual EPA approach of developing "model" mill data and then generalizing to the universe of mills, we asked McCubbin [1988] to develop a data base containing the bleaching process and estimated chemical consumption for 138 bleaching lines in 103 mills. He could not find information about one mill. For each bleaching line, he calculated the TOX discharge with the current configuration. His calculations, based on a modification of Germgard's equation, indicate that all mills, except those few mills with ocean discharges and no biological treatment, would be in compliance with a regulatory limit of nine pounds of TOX per ton of air-dried pulp.

McCubbin first estimated the costs of reducing TOX discharge for the 65 bleach lines at the 55 mills discharging over five pounds of TOX per air-dried ton. He assumed that the 55 mills would use high chlorine dioxide substitution, which usually does not require a large capital investment. For each mill, he estimated where necessary the cost of: (1) converting the caustic extraction stage to use elemental oxygen at atmospheric pressure to reduce total chlorine input, (2) upgrading or replacing existing chlorine dioxide production facilities, and (3) installing new chlorine dioxide production facilities where none exist. He calculated an average capital cost per mill of $3.2 million, with an increase of $475,000 per year in chemical costs. (See Appendix 8-A for cost and effluent data for individual mills.) The average annual costs would be approximately $1.1 million. The total capital costs for industry compliance would be about $175 million, the total chemical costs would be about $26 million, and annual costs would be about $60 million.[4] The industry would spend on average $3 per air-dried ton and would reduce TOX discharge from 195 tons per day to 108 tons per day. The average TOX discharge for all mills would fall to three pounds per air-dried ton.[5]

McCubbin also estimated the costs of reducing TOX discharge for 120 bleach lines at 94 mills that discharge over three pounds of TOX per air-dried ton. He assumed that each bleach line would install, where necessary, an oxygen delignification stage upstream of the bleach line as well as make the process modifications required to meet the 5-pound TOX limit. He did not take cost credits for energy recovery from the oxygen stage, nor did he allow any costs associated with the increased load in recovery boilers. For the average mill, the capital investment to meet the

3-pound limit would be $10.5 million, with a savings in chemical costs of $520,000. The annual costs would be approximately $1.2 million. The total capital costs for the industry would be about $1.0 billion, the total chemical cost savings would be about $49 million, and annual costs would be about $110 million. The mills complying with the limit would spend on average $7 per air-dried ton.[6] The industry would reduce its estimated 1988 TOX discharge from 195 tons per day to 54 tons per day. The average TOX discharge for all mills would fall to 1.5 pounds per day per air-dried ton.[7]

Dioxins

Much of the recent interest in reducing organochlorine discharge from pulp mills has centered on the presence of extremely small quantities of dioxin, particularly 2378-TCDD, among the hundreds of compounds that make up the total organochlorine discharge. There is no scientific basis for assuming that reduction of organochlorine discharges will result in corresponding reductions in 2378-TCDD discharges. About 200 pounds of chlorine are required per ton of pulp processed by a typical kraft bleach plant. Of this, about 10 pounds are discharged in the form of chlorinated organic compounds, and most of the rest are discharged in the form of relatively innocuous chlorides. Under 1 ten-millionth of the organochlorine is made up of 2378-TCDD.

Fortunately, the limited data currently available [Voss et al. 1988] indicate that substantial or complete elimination of detectable amounts of 2378-TCDD probably can be achieved by proper selection of commercial defoamer products, combined with high chlorine dioxide substitution and/or oxygen delignification. This conclusion must be considered tentative, however, until some of the extensive investigations currently under way are completed. The average TOX discharge for all mills would fall to three pounds per air-dried ton.

Regulatory Approaches

Following our usual pattern, we examined alternative regulatory approaches for reducing TOX effluents from bleached kraft mills. We wanted to determine if we could: (1) characterize the benefits from alternative toxic pollutant policies (as we did for conventional pollutants), and (2) empirically identify one policy as preferable on the grounds of the net benefits criterion. Only if we could adequately characterize the risk reductions resulting from alternative policies could we then distinguish among regulatory approaches on the basis of generating net benefits.

Technology Approach. A technology approach would require all mills to meet a TOX discharge limitation (Table 8-2). We found that 55 of the 103 mills discharge more than five pounds of TOX per air-dried ton. They could reduce their discharge below five pounds using 50 percent substitution of chlorine dioxide at an annual cost of approximately $60 million. We found that 94 of the 103 mills discharge more than three pounds of TOX per air-dried ton. They could reduce their discharge

Table 8-2. Evaluation of Alternative Approaches to Regulating
Dioxin as a Toxic Water Pollutant ($1984 10^6)

Regulatory Approach	Bleach Lines / Mills Controlled / Mills Investigated	Annual Costs[a]	Qualitative Benefit Assessment
Technology (5 lb. limit)	65 / 55 / 103	$60	This standard could reduce potential cancer incidence by 50 percent and some violations of the criterion for aquatic life protection.
Technology (3 lb. limit)	120 / 94 / 103	$110	This standard could reduce potential cancer incidence by 75 percent and most violations of the criterion for aquatic life protection. It would not eliminate violations of criterion for human health protection.
Ambient	39 / 30 / 103	$40	This standard could eliminate most quality violations of the water criterion for aquatic life protection.
Benefits	N/A	N/A	N/A

N/A = No estimate available.
a. Based on 1984 capacity.

air-dried ton. They could reduce their discharge below three pounds using oxygen delignification in combination with substitution of chlorine dioxide at an annual cost of approximately $110 million.

Given the uncertainties about how 5-pound and 3-pound TOX limitations would reduce dioxin concentrations in the receiving waters downstream from pulp-producing mills, we hesitated to characterize the results from such limitations. Also, we questioned our own estimates of cancer incidence from eating dioxin-contaminated fish. Furthermore, we do not know how fish contamination would change as a result of reducing TOX discharges. With these caveats, we speculate that the more stringent TOX limit of three pounds would reduce the cancer incidence by 75 percent (paralleling the 75 percent reduction in TOX discharge) and would diminish to some extent the adverse effects on fish and wildlife.

Ambient Approach. An ambient approach would focus on reducing violations of the ambient water quality standard for dioxin. On the basis of our water quality analysis, we think that the discharges from most mills are resulting in violations of the ambient water quality standard for dioxin based on human health protection and that the discharges from some mills are resulting in violations of an ambient standard that would protect aquatic life. We chose to base our evaluation of an ambient-based policy on the aquatic life standard because it allowed us to identify a subset of the 103 mills as potentially causing more serious health risks as well as ecological risks.

We started with the 33 mills identified by the water quality simulation as causing violations of the aquatic life standard under average flow/high dioxin discharge conditions. Because the dioxin concentrations at individual mills are based on the "5-mill" study rather than mill-specific concentration data, we examined the mill-specific data on TOX discharge. We first eliminated three mills with TOX discharges less than three pounds per ton. The remaining 30 mills (39 bleach lines) could meet a TOX discharge limit of three pounds per ton at an annual cost of $39 million, which is approximately 40 percent of the costs of all 94 mills complying with a 3-pound limit. We next eliminated 78 mills with TOX discharges less than seven pounds per ton (the maximum is 7.8 pounds per ton). The remaining 20 mills (31 bleach lines) could meet a TOX discharge limit of three pounds per ton at an annual cost of $31 million. We took the 30-mill effluent reduction/cost estimate as the basis for a conservative ambient-based policy because it was not significantly different from a 20-mill effluent reduction. We could not simulate how many of the 30 mills would comply with an aquatic life protection standard, if they met the 3-pound TOX limit. We did not know whether mill-specific dioxin discharges would decrease with reductions in TOX. If the 3-pound discharge limit reduces dioxin from high to low concentrations (Figure 8-2), then virtually all mills would comply with the aquatic life standard.

Benefits Approach. A benefits approach would require process modifications at only those plants where the reductions in health and ecological risks would be comparable to the costs. Given the lack of plant-specific information, we could not evaluate a benefits approach.

Findings -- Dioxin

Dioxin discharges from bleached kraft mills appear to be a pervasive and significant problem. Approximately one-half of the mills are discharging more than five pounds of TOX per ton of production. Dioxin in the TOX discharge is causing violations of the EPA water quality standard for human health protection at most mills and is causing violations of a suggested water quality standard for aquatic life protection at 10 to 40 mills. And, if EPA's cancer potency and fish consumption calculations provide a reasonable basis for estimating cancer incidence, then 0.6 to 6.0 excess annual cancer incidents result from eating contaminated fish.

Reduction of the dioxin discharge is technologically feasible and relatively inexpensive -- $3 to $7 per ton with 1984 chemical costs ($3 to $4 per ton with 1988 chemical costs). The industry sold market-bleached pulp in 1988 for $600 to $700 per ton and spent between $15 to $20 per ton to meet existing environmental regulations. The additional expenditure would not be unreasonable in light of the current situation.

Although we think that curtailing dioxin discharges would generate net benefits, we could not empirically conclude which of the three approaches would be significantly superior in producing this result. The absence of mill- and site-specific data prevented us from fairly characterizing and ranking the three policies. In situations that combine pervasive and significant risk and reasonable pollution reduction costs, however, a feasible technology-forcing approach, such as a 3-pound TOX discharge limit, would be appropriate. Mill- and site-specific analyses are needed to determine quantitatively whether this regulatory policy would generate net benefits.

CHLOROFORM -- AIR POLLUTANT

Chloroform is a well-known chemical that for years was used in high concentrations as a human anesthetic. At very high chronic levels (>100 ppm in drinking water), unequivocal topical effects occur in rat and mice livers.

Emissions

The pulp and paper industry forms the largest industrial chloroform emissions category, accounting for approximately 45 percent of chloroform emissions in the United States in 1982 [EPA. 1985(d)]. Chloroform is produced indirectly in the process water during the bleaching of wood pulp in approximately 125 mills. Chloroform formed in process water subsequently evaporates to the atmosphere during both the treatment of process wastewater and after discharge from treatment into receiving waters.

Health and Ecological Effects

EPA's Office of Health and Environmental Assessment reviewed the scientific literature to assess the health and ecological effects of exposure to chloroform [EPA. 1985(c)]. Although it found sufficient evidence to conclude that chloroform is carcinogenic in experimental animals, confirming evidence from human studies is limited to epidemiological studies of ingestion of chlorinated drinking water (chlorinated drinking water contains chloroform as well as other trihalomethanes). Although the results of these studies suggest increased incidence of cancer in humans, chloroform could not be implicated as the sole causative agent. Therefore, the available direct evidence for chloroform carcinogenicity in humans is inadequate to assess its carcinogenic potential. EPA concluded that the evidence for carcinogenicity would place chloroform into group B2 (i.e., probably carcinogenic to humans).

EPA also concluded that significant ecological effects are not expected because chloroform is quite volatile and does not accumulate in terrestrial or aquatic environments. Acute effects on aquatic life can occur in the vicinity of major chloroform spills, but significant chronic effects from long-term exposure to low ambient levels is unlikely. The limited bioconcentration and short biological half-life of chloroform suggest that its residues would not be an environmental hazard to consumers of fish.

Risks to Public Health

To assess the risk of contracting cancer from exposure to ambient concentrations of chloroform, EPA derived an upper-limit unit risk estimate of carcinogenic potency from the available health information.[8] For the purposes of our analysis, we accepted EPA's linear, no-threshold assumption for assessing the risks from exposure to atmospheric chloroform. By using the information from rodent oral exposure assays and an estimate of chloroform absorption through the human lung, EPA's Carcinogen Assessment Group derived a unit risk factor associated with one ug/m^3 of chloroform in air of 2.3 x 10^{-5}. This means that the additional probability of an individual's contracting cancer due to continuous 70-year lifetime exposure to one microgram of chloroform per cubic meter of air is 2.3 chances in 100,000.

We think that the unit risk factor adopted by EPA's Office of Air Quality Planning and Standards is at least one, if not two, orders of magnitude too high. Other organizational units within EPA, the Office of Drinking Water and the Office of Solid Waste, would agree that the unit risk factor is too high by at least one order of magnitude [FR. 1988]. Later in this chapter we use the Office of Solid Waste's unit risk factor in critiquing EPA's designation of chloroform as a hazardous waste constituent generated by the pulp and paper industry. We provide a critique of the risk assessment in Appendix 8-B, which concludes that the risk assessment could be a factor of 500 times too high, even without questioning the assumption of a linear non-threshold model.

EPA estimated public exposure to chloroform emissions from pulp and paper mills using the Human Exposure Model [EPA. 1986]. To estimate human exposure, the Human Exposure Model uses information on meteorological conditions and population distribution. It combines a Gaussian distribution model, emission release parameters, and meteorological data to estimate the concentration of a pollutant in the ambient air at different points within a radius of the source. Based on the radial distance from the source, out to 50 kilometers in this case, it numerically combines estimated pollutant concentrations with population data to produce quantitative expressions of public exposure to the pollutant.

EPA combined the upper-bound unit risk factor with the exposure information from the Human Exposure Model to estimate the quantitative risk that chloroform poses to public health [EPA. 1985(b)]. The quantitative health risks are expressed as maximum individual risk and annual aggregate incidence. The maximum individual risk is the added lifetime probability of the most exposed individual contracting cancer as a result of exposure to chloroform in the ambient air. The annual aggregate incidence is an estimate of the excess cancer cases every year (nationwide) attributable to exposure to ambient air concentrations of chloroform.

A summary of the public health risk estimates from EPA's 1985 analysis for 125 pulp and paper mills is presented in Table 8-3. The maximum individual risk is 1.1×10^{-2} and the aggregate cancer incidence is 0.80 cases per year. A 1989 analysis of exposure using the Human Exposure Model suggests that the maximum individual risk is 2.0×10^{-2} and the aggregate cancer incidence is between 1.5 and 2.0 cases per year at pulp mills [Radian. 1989]. More recent site-specific modeling, however, suggests that the 1985 estimates are more accurate [Vandenberg. 1990]. Consequently, we retained the original individual risk and cancer incidence estimates for our analysis.

Table 8-3. Summary of Pulp Mills with Chloroform as a Toxic Air Pollutant

Sub-Category	Number of Mills	People Exposed[a] (10^3)	Highest Individual Risk	Cancer Incidence (cs/yr)
Soda & Kraft Fine	19	7,600	2.8×10^{-3}	.12
Papergrade Sulfite	12	4,700	1.1×10^{-2}	.30
BCT Bleached Kraft	9	1,600	5.5×10^{-3}	.05
De-ink Fine	4	4,700	9.2×10^{-4}	.05
De-ink Tissue	16	17,800	2.2×10^{-4}	.02
Dissolving Kraft Pulp	3	200	2.3×10^{-3}	.007
Dissolving Sulfite Pulp	6	700	2.2×10^{-6}	.007
Market Bleached Kraft	12	3,100	4.6×10^{-3}	.07
Misc. Integrated	44	17,600	2.0×10^{-3}	.17
TOTAL	125	57,900		.80

a. People within 50 km.
Source: "Chloroform Exposure and Risk Assessment" [EPA. 1985(b)].

Emission Reduction Technology and Costs

The use of calcium or sodium hypochlorite for bleaching pulp is recognized as the major source of chloroform in the pulp and paper industry. Over 90 percent of the chloroform concentrations is formed in the hypochlorite bleach stages. Effluent from the bleaching operation typically is discharged to on-site pulp mill wastewater treatment plants. The influent and effluent chloroform shows a sharp decrease in wastewater chloroform concentrations during treatment, a trend attributed to evaporation.

Because these wastewater treatment plants at pulp mills treat large quantities of wastewater and often use several acres of stabilization ponds, capturing the chloroform from treatment plants is not feasible by any available technology. Emissions of chloroform from pulp and paper mills must be controlled by modifying the bleaching process to reduce or prevent chloroform formation. Process modification consists of replacing hypochlorite with chlorine dioxide, which produces virtually no chloroform, or with molecular oxygen, which does not react to form chloroform.

The replacement bleach sequence should be developed on a case-by-case basis because pulp mills vary in physical design and type of end product. Several different pulp mills use a few common sequences, however. For its analysis, EPA categorized pulp mills based on end product and bleach sequence. EPA identified the most common bleach sequences using hypochlorite and grouped them according to their manufacturing capacity. Each category is represented by a "model" pulp mill characterized by a bleach sequence (using hypochlorite) and production capacity. EPA contacted manufacturers of pulp bleaching equipment to obtain information on feasible substitute sequences using chlorine dioxide and oxygen. Appendix 8-C summarizes the data about the model mills, their substitute bleach sequences, and includes a description of the proposed modifications.

Total chloroform emissions, based on influent chloroform concentration in mills in all nine subcategories that bleach with chlorine or chlorine-containing compounds, are approximately 4,280 tons per year. Total annual chloroform emissions from all mills in the five subcategories for which EPA considered process modifications are estimated to be 3,670 tons per year, 86 percent of the total.

EPA assumed that four categories of pulp mills -- dissolving sulfite pulp, dissolving kraft pulp, de-ink tissue, and de-ink fine paper -- would continue to use hypochlorite because they could not make the necessary process modifications. Thus, no emission reductions or costs are presented for these categories. EPA also assumed that pulp mills in the other five categories would reduce chloroform emissions by 92 percent as the result of process modification. The estimated chloroform emission reductions at each mill range from five tons per year to 170 tons per year.

EPA estimated control costs for eight model mills based on the modifications. The model mill control costs ranged from a net savings of $25,000 per year (due to reduced chemical costs) to a net cost of $2.2 million per year. The control costs for existing pulp mills were drawn from the costs for the model mill that most closely matched the bleach sequence and size of the existing mill. Where the model mills did not represent an existing mill, the costs for that mill were computed separately. Estimated annualized control costs for existing mills ranged from a net savings of $25,000 to a net cost of $5.55 million per year.

For the analysis presented in this chapter, we re-estimated the control costs for the eight model mills, because we believe that EPA's calculations are incorrect. Appendix 8-C compares our model-mill costs with EPA's model-mill costs. Essentially, we lowered the capital recovery factor, the purchase cost of chemicals, and the amount of chemical usage.

Regulatory Approaches

Following our usual pattern, we examined alternative regulatory approaches for reducing chloroform air emissions from bleached kraft mills. We wanted to determine if we could: (1) characterize the benefits from alternative toxic pollutant policies (as we did for criteria pollutants), and (2) empirically identify one policy as preferable on the basis of the net benefits criterion. Only if we could adequately characterize the risk reductions resulting from alternative policies could we distinguish among regulatory approaches on the basis of generating net benefits.

Technology Approach. A technology approach to regulating chloroform would require, where possible, process modifications for all 125 mills. EPA assessed that four of the nine production subcategories (29 mills) could not modify their production processes and would continue using hypochlorite. EPA found that process modifications were feasible only at 70 of the 96 mills in the other five categories. Thus, risk reductions and costs were calculated for only those 70 mills. (See Appendix 8-D for individual mill cost estimates.)

Application of process modifications, assuming 92 percent emission reductions at the 70 mills, would result in a 0.63 annual cancer incidence reduction at a cost of $44.3 million (Table 8-4).[9] This 80 percent reduction in the annual incidence would cost $70.0 million per case avoided.

Risk Reduction Approach. For this pollutant, analyzing an ambient approach would not be feasible because there is no ambient air standard for chloroform. In its place, we analyzed a risk reduction approach that would focus on reducing the higher maximum individual risks from chloroform around pulp mills. A focus on reducing the higher maximum individual risks would require process modification at only those plants whose emissions result in a general population risk of 10^{-3} or higher (Table 8-4). In this case, EPA would require approximately 33 plants to modify their bleaching sequence at an annual cost of $18.5 million, or approximately 40 percent of the annual costs for all 70 mills. Cancer incidence reduced would be 0.42 cases annually, or approximately 66 percent of the cancer incidence reduction with a technology-based policy. The cost per case avoided would be $50.0 million, rather than the $70.0 million for the technology approach.

Table 8-4. Evaluation of Alternative Approaches to Regulating Chloroform as a Toxic Air Pollutant ($1984 10^6)

Approach	Mills Controlled / Mills with Emissions	Annual Costs	Cancer Incidence Reduction/Total Cancer Incidence	Cost/Cancer Incidence Reduced
Technology	70/125	$44.3	.63 / .80	$70
Risk (10^{-3})	33/125	$18.5	.42 / .80	$50
Risk (10^{-2})	1/125	$1.1	.14 / .80	$ 8
Benefits	17/125	($3.3)	.29 / .80	$ 0

Another variation of a risk reduction approach would require process modification at only those plants whose emissions result in the highest individual risk, which in this case is 10^{-2}. In this case, EPA would require only one plant to modify its bleaching sequence at a cost of $1.1 million, or approximately two percent of the annual costs for all 70 mills. Cancer incidence reduced would be 0.14 cases annually, or approximately 22 percent of the cancer incidence reduction with uniform application of process modifications. The cost per case avoided would be $8.0 million, rather than $70.0 million.

Benefits Approach.   A benefits approach would require process modifications at only those plants where the cost of incidence avoided was $10 million or less.[10] It would require approximately 17 plants to modify their bleaching sequence at an annual cost saving of $3.3 million. (There would be pollution control costs at two of the mills and positive cost savings at 15 mills.) Cancer incidence would be reduced by 0.29 cases annually, or approximately 40 percent of the cancer incidence reduction with a technology-based policy. The cost per case avoided would be zero in the aggregate, rather than $70.0 million under the technology approach.

Findings -- Chloroform as a Hazardous Air Pollutant

We found that we could adequately describe the benefits resulting from alternative regulatory policies and could identify a benefits-based policy as the most appropriate one for generating net benefits. A benefits-based policy would reduce approximately 40 percent of the population risk at no cost, compared to a $44.3 million cost for a technology-based policy. Emission reductions at the 17 mills that would be regulated under a benefits-based policy would protect all individuals with a maximum risk of 10^{-2} (33 people) and slightly more than one-half of those individuals with a maximum risk of 10^{-3} (90 out of 155 people). At 15 of the 17 mills, the policy would result in positive savings to the mills while at the same time reducing risk in the surrounding communities.

CHLOROFORM -- HAZARDOUS WASTE

The 1984 amendments to the Resource Conservation and Recovery Act require EPA to revise its process for identifying hazardous wastes that are toxic.   Under this act, EPA has proposed a rule that will require generators of hazardous wastes to determine whether their wastes contain concentrations of 25 organic chemicals that exceed regulatory levels. Currently, generators must make this determination for 14 toxic contaminants (eight metals, four pesticides, and two herbicides) using the Extraction Procedure test. The proposed rule would replace that test with the Toxicity Characteristic Leaching Procedure. Failing this test for any of these 25 constituents implies that the generator must manage the wastes under the hazardous waste regulations (Subtitle C) of the Resource Conservation and Recovery Act.

This regulatory effort reflects a predominant concern with possible groundwater contamination that would result in exposure for humans through drinking water at a wellhead. To estimate exposure at the wellhead, EPA modeled the process through which toxic constituents in the wastes leave the wastewater impoundment, enter the ground water, and move to a wellhead. For the toxicity characteristic regulation, EPA applied a probabilistic groundwater fate and transport model to estimate toxic concentrations at potential drinking water sources.[11]

The limit on the constituent's concentration in the waste can be calculated once a constituent's toxicity reference level at the wellhead is chosen. The chronic toxicity reference level is the health-based threshold for a contaminant's concentration in drinking water. For each constituent, it will be either the national interim primary drinking water standard or the maximum contaminant level where available. Otherwise, the risk-specific dose for a carcinogen, or reference dose for a noncarcinogen, will be used.

Groundwater contamination has other potential negative impacts besides increases in health risks. There can be losses in resource value that result from contamination. Contaminated water may be considered unusable for drinking, cooking, or bathing if the level of contamination makes the water unpleasant. Taste and odor thresholds for each constituent determine whether there has been significant resource damage. These thresholds typically exceed the toxicity levels.

Pulp and Paper Wastes

Pulp and paper is one of the major industries to be affected by the toxicity characteristic rule. The industry manages very large volumes of wastewaters in surface impoundments, many of which would not meet Subtitle C requirements. A survey of industrial establishments estimated that the pulp and paper industry managed approximately 2,020 million metric tons in surface impoundments, which was nearly one-third of the total wastes managed in surface impoundments among the 17 industries surveyed [Westat, Inc. 1987]. Current estimates of wastewaters that will actually fail the toxicity characteristic test are smaller but still substantial.

Concentrations of chloroform in the waste streams of most of the industry's sectors exceed the regulatory level (6.0 mg/l) and appear to be the major source of health risks and resource damages estimated from the industry's wastes.[12] Tetrachloroethylene appears to be a potential source of risks and resource damages in the "De-inking - Fine Tissue Paper Production" sector. Trichloroethylene is also present in de-inking wastes.[13] The presence of 2,4,6-trichlorophenol in the waste streams of various industry sectors may generate minor resource damages. Benzene is present in hardboard manufacturing wastewaters, but concentrations do not seem to exceed the regulatory level [ICF, Inc. 1988].

Risks to Public Health

Cancer Risks. Private wells are a primary source of exposure to groundwater contamination because it is less likely that their water will be tested and treated than is the case with public wells. Using Census data on well use density by zip codes throughout the United States, we derived indications of the differences in private well exposure for the mills in this study. These mills exhibited sufficient variation in private well use density to raise concerns about an approach that requires uniform compliance from all mills throughout the country. These densities are notably smaller than the average well-using density of 1.6 users per acre assumed in the Toxicity Characteristic Regulatory Impact Analysis [ICF, Inc. 1989].

Based on the private well densities, the estimated population cancer risks from 48 of 62 mills where data were available are extremely small (Table 8-5). The highest estimate for an individual mill, using an upper-bound estimate for the dispersion of a contaminated plume and assuming no volatilization of chloroform, is 2.0 E^{-6} of a cancer incidence per year. The estimate for 48 mills is 3.6 E^{-5} of a cancer incidence per year. Although larger when public well users are included in the facility-specific well-user densities, the estimated risks are still negligible. The estimate for cancer incidence combining private and public well users is 4.5 E^{-5} of a cancer incidence per year.[14]

Resource Damages. ICF, Inc. [1989] calculated resource damages on the basis of providing alternative water supplies to contaminated private wells. Because the fixed costs of providing an alternate water supply constitute a large portion of the costs, the magnitude of the damages depends more on the size of the contaminated plume from a mill than on the size of the exposed population. The lower-bound estimate of plume size assumed a wastewater impoundment of 6,400 square meters, and the upper-bound estimate assumed a wastewater impoundment of 4.3 million square meters. This range encompasses the sizes of wastewater impoundments at a majority of the mills.

Table 8-5. Summary Data on Pulp Mills with Potentially Hazardous Waste Pollutants

Subcategory	Number of Mills Analyzed / Number of Mills Potentially Regulated[a]	Total Households Exposed (10^3)	Higher Well Densities[a]	Cancer Incidence
Papergrade Sulfite	8 / 10	22 / 27.5	1	$4.4E^{-6}$
De-inking - Fine Tissue Paper	11 / 15	48 / 68	3	$1.4E^{-5}$
Bleached Kraft	15 / 22	11 / 22	2	$6.5E^{-6}$
Alkaline - Soda and Fine Bleached	14 / 15	26 / 38	2	$1.1E^{-5}$
Total	48 / 62	110 / 155	8	$3.6E^{-5}$

a. Number of mills with well-using density ≥ 0.1 users/acre.

For the purpose of this analysis, we took the upper-bound estimate, which results in a greater plume size and therefore higher costs for resource damages avoided. However, we question the cost estimate for alternative water supplies because it includes both capital and operation and maintenance costs. If private and public well users are now bearing operation and maintenance costs, then only operation and maintenance costs above those now being incurred should be included in the resource damage estimate. Consequently, we calculated a cost range for resource damage, with the lower bound reflecting no additional operation and maintenance costs, and the upper bound including full operation and maintenance costs.

Effluent Reduction Technology and Costs

The compliance costs for this rule will depend on how wastewaters and non-wastewaters (sludge) are managed in the baseline. Large-volume wastewaters that are not currently regulated by Subtitle C, but that fail the toxicity characteristic test, are typically managed in surface impoundments. Under the Resource Conservation and Recovery Act requirements, their surface impoundments must be double-lined and must meet other design and operation standards.

Rather than retrofit currently unlined surface impoundments, wastewater generators are expected to manage these wastes in tanks because this option is less expensive. Certain tanks used for wastewater treatment before discharge to a surface water are exempted from Resource Conservation and Recovery Act permit requirements if the facility has a National Pollutant Discharge Elimination System permit. Furthermore, the toxicity characteristic test applies to wastes "at the point of generation," which has been interpreted to mean that when wastes no longer exhibit the characteristic in a treatment train, Subtitle C regulations no longer apply. Consequently, the move from management in a surface impoundment to tanks may be limited to initial impoundments of wastewater, thereby limiting the retrofitting necessary for compliance.

The management of sludges will probably be altered as a result of this rule. Sludges not previously regulated by Subtitle C have been primarily managed by land disposal. Land disposal is still an option under Subtitle C, but the stringent requirements that have been imposed on land disposal under the Resource Conservation and Recovery Act dramatically increase the costs of managing sludge. Furthermore, these costs are likely to increase, as the result of increased requirements for land disposal, including treatment requirements for land-disposed wastes defined by the toxicity characteristic.[15]

For the analysis presented in this chapter, we took the unit cost estimates from the Toxicity Characteristic Regulatory Impact Analysis [ICF, Inc. 1989]. We based the individual mill cost estimates on wastewater flow data from EPA's Industrial Facilities Discharge file. The annual costs of compliance with the proposed rule would be $30.3 million.[16]

Regulatory Approaches

Following our usual pattern, we examined alternative regulatory approaches for managing toxic wastes, primarily those associated with chloroform, from pulp mills. We wanted to determine if we could: (1) characterize the benefits from alternative toxic waste management policies, and (2) empirically identify one policy as preferable for generating net benefits. Only if we could adequately describe the benefits of alternative policies could we distinguish among regulatory approaches on the basis of economic efficiency.

Technology Approach. A technology approach to regulating toxic wastes, primarily chloroform, would require all 62 mills to manage their wastewaters and sludge. For the 48 mills where we had both cost and benefit information, the annual costs of toxic waste management would be approximately $30.3 million (Table 8-6).

Table 8-6. Evaluation of Alternative Approaches to Regulating Chloroform as a Hazardous Waste ($1984 10^6)

Approach	Mills Controlled / Mills Analyzed / Mills with TC Wastes	Annual Costs[a]	Incidence Reduction / Total Cancer Incidence	Cost/Cancer Incidence Reduced	Resource Damages Avoided w/o O&M	w/
Technology[b]	48/48/62	$30.3	$3.0E^{-5}/3.6E^{-5}$	$800,000	$1.1	$3.2
Exposure[c]	7/48/62	$5.0	$2.1E^{-5}/3.6E^{-5}$	$200,000	$0.19	$0.5
Benefits[d]	8/48/62	$0.2	$4.4E^{-6}/3.6E^{-5}$	$50,000	$0.14	$0.4

a. Based on 1984 capacity.
b. Includes all four subcategories.
c. Includes mills with ≥ 0.1 well users per acre only.
d. Includes papergrade sulfite subcategory only.

The benefits from this regulatory approach can be characterized as -reduction of cancer incidence or as resource damage avoided. Cancer incidence reduced would be very small ($3.6E^{-5}$), and the cost per incidence reduced would be $800 billion. Alternatively, the resource damages avoided -- i.e., the costs of alternative water supplies -- would be between $1.1 million and $3.5 million.

Exposure Reduction Approach. For toxic wastes from pulp mills, analyzing either an ambient-based or a risk-reduction approach was not feasible. We could not evaluate an ambient approach because EPA does not know either the level of toxic constituents in the wastes of the 48 mills or their fate in the groundwater surrounding their mills.[17] Neither could we evaluate a risk reduction approach, a second best ambient approach, because EPA does not know the concentration of toxic constituents in the groundwater surrounding the mills nor the actual number of private or public wells using the groundwater surrounding the mills.

In their place, we analyzed an exposure-reduction approach that would focus on protecting groundwater that potentially supplies a significant number of private wells. For this analysis we assumed that "significant" health risk would occur in areas where the well-using population density was greater than 0.1 well-using people per acre. In this case, EPA would require approximately seven mills to manage their toxic wastes at an annual cost of $5.0 million, or approximately 15 percent of the annual costs for all 47 mills. Cancer incidence reduced would be very small (2.1 E^{-5}), or approximately 60 percent of the cancer incidence reduction with a technology-based policy. The cost per incidence avoided would be $200 billion. Alternatively, the resource damages avoided would be between $0.2 million and $0.5 million.

Benefits Approach. A benefits approach would require toxic waste management at only those mills where there is the potential for generating net benefits. In this case, we could not identify a subset of the 48 mills where the benefits would exceed costs because of the lack of mill-specific benefits data. Instead, we examined the one subcategory of mills, papergrade sulfite, where there is the possibility of reducing relatively higher risks. The annual costs of toxic waste management ($0.2 million) are relatively low for this subcategory because only the wastewaters and not the sludges are considered toxic. Moreover, the potential benefits might be comparable to the costs, based on estimated resource damages avoided by private well users ranging from $0.1 million to $0.5 million.

We doubt, however, that the regulation of this one subcategory of mills would meet the net benefits criterion. The cost per cancer incidence reduced is $50 billion, which clearly falls outside EPA's guidelines for cost-effective risk reductions [EPA. 1983]. Also, the population around six of the seven papergrade sulfite mills uses or has access to community water supply systems. In these situations, the resource damages avoided are much lower than those that are estimated in the ICF analyses that are displayed in Table 8-6.

Findings -- Chloroform as a Hazardous Waste

We found that we could only generally characterize the benefits resulting from alternative regulatory policies and could not identify a policy with the potential for generating net benefits. A potential benefits-based approach would reduce approximately 35 percent of the population risk at an annual cost of $0.2 million. We are not convinced, however, as we were for the benefits-based policy for chloroform as a toxic air pollutant, that the benefits of this policy would exceed the costs.

FINDINGS

Limitations in data and analytical techniques, as well as in the resources available to evaluate the environmental risks from the pollutants of concern, constrained our analyses of toxic pollutants. Our findings suggest that regulating some toxic pollutants -- dioxin as a toxic water pollutant and chloroform as a hazardous air pollutant -- would produce net benefits. While only a benefits approach would reduce unreasonable risks for chloroform as a hazardous air pollutant, any of the three approaches would be efficient for dioxin in wastewater. Our findings also suggest that any approach for reducing the other toxic pollutant -- chloroform as a hazardous waste constituent -- would not result in net benefits. The costs of regulating chloroform as a hazardous waste appear to be disproportionate to the human health risks. If EPA were to regulate chloroform as a hazardous air pollutant, however, most of the chloroform in the solid waste from pulp and paper mills would disappear.

Notes

1. Total organochlorines include chlorophenols, chloroform, chlorinated acetic acids, chlorinated dimelhysulfurans and chlorinated dioxin and similar compounds.

2. Organic chlorine = k* (C + H/2 + D/5) kg/ton pulp

> where: k is constant in the range 0.07 to 0.11 if C, H, and D are in the units of kilogram per metric ton pulp and the organic chlorines are expressed according to the TOCl analysis procedure [Sjostrom et al. 1982];
>
> C is total elemental chlorine charge;

H is the hypochlorite charge, as equivalent chlorine
(= 1.5 * NaOCl);

D is chlorine dioxide charge as equivalent chlorine
(= 2.63 * ClO_2).

Earlier literature proposed values from 0.05 to 0.07 for k, but personal communication with Germgard by Bonsor et al. [1988] and analysis of various reported data suggest modifications to the equation. Where APHA TOX data are to be calculated, a multiplier of approximately 1.3 must be used. More recent data [Cook. 1988; Hall et al. 1988], as well as extensive discussions with laboratories that have not yet published data from their intensive work in early 1989, suggest that TOX can be calculated as 0.1 times the chlorine charge, and that neither chlorine dioxide nor sodium hypochlorite bleaching agents contribute to organochlorine discharges.

3. The representativeness of this range for all 74 mills is supported by a more recent EPA survey of dioxin discharges from 103 mills [EPA. 1989(a)]

4. The annual cost estimate is very sensitive to chemical costs, especially the ratio between the cost of caustic/chlorine and the cost of sodium chlorate. Using 1988 rather than 1984 cost estimates, the costs of chemicals for all mills would be $12 million rather than $26 million, and the resulting annual costs would be approximately $40 million rather than $60 million.

5. This reduction is greater than might be expected because some mills are already below that limit, and because it is probable that the mills installing new technology would use the equipment to its fullest to respond to consumer pressure to reduce dioxin in paper.

6. Here, too, the annual cost estimate is very sensitive to chemical costs. Using 1988 rather than 1984 cost estimates, savings in chemical costs for all mills would be $132 million, rather than $49 million, and the resulting annual costs would be approximately $35 million, rather than $110 million. As a result, the industry would spend on average $4, rather than $7, per air-dried ton to meet a 3-pound TOX discharge limit.

7. This reduction is greater than might be expected, because some mills are already below that limit and because the mills installing new technology would use the equipment to its fullest.

8. The unit risk factor is an estimate of the additional probability that an individual will die from cancer resulting from continuous lifetime exposure to 1 microgram of chloroform per cubic meter of inspired air (assuming a 70-year lifespan).

9. The EPA model mill costs and estimates of pulping capacity, rather than our estimates, result in an annual cost for process modification at the 77 mills of $140 million. In this case, the 80 percent reduction in annual incidence would cost $217 million per incidence avoided.

10. Ten million dollars is slightly greater than the upper-bound monetary value that EPA's Guidelines ascribe to reductions in cancer incidence [EPA. 1983].

11. The model employs a Monte Carlo simulation, which uses distributions of the parameters of the fate and transport system as a means of incorporating the variation in these parameters that is expected at sites where toxic wastes are managed. At the time of this writing, EPA is expected to retain a dilution attenuation factor of 100 for all constituents [FR. 1988].

12. The regulatory level is a function of the human health standard of 0.06 mg/l and the proposed dilution/attenuation factor of 100 [FR. 1988]. A concentration above 6.0 mg/l in the wastewater and sludge of pulp and paper mills would mean that the mill would be subject to regulation.

13. Although tetrachloroethylene presents greater risks, the presence of trichloroethylene has a greater effect on compliance costs.

14. If volatilization of chloroform is incorporated, all cancer incidence estimates fall by 70 percent.

15. There are no restrictions on which wastes can be disposed of in land units with Subtitle C permits. However, the Resource Conservation and Recovery Act requires EPA to establish treatment standards for land-disposed wastes defined by the toxicity characteristic by May 8, 1990.

16. The annual costs (and wastewater flows) for the four subcategories would be: $0.2 million (115,000 gallons per day) for papergrade sulfite, $3.7 million (43,000 gallons per day) for deinking-fine tissue paper, $14.1 million (435,000 gallons per day) for bleached kraft, and $12.3 million (365,000 gallons per day) for alkaline-soda and fine bleached.

17. The technology approach simulated the fate of generic concentrations of toxic constituents in mill wastes in hypothetical groundwater environments. These simulations are not adequate for estimating site-specific risks.

References

American Paper Institute (API). 1968. *Paper Paperboard Capacity.* New York, NY.

_____. 1985. *Paper Paperboard Woodpulp Capacity.* New York, NY.

_____. 1989. *Paper Paperboard Woodpulp Capacity.* New York, NY.

APHA. 1985. "Standard Methods for the Examination of Water and Wastewater," 16th edition. American Public Health Association, Washington, DC.

Arthur D. Little, Inc. 1987. "Exposure and Risk Assessment of Dioxin in Bleached Kraft Paper Products." Report to the Office of Water Regulations and Standards. U.S. Environmental Protection Agency, Washington, DC.

Bonsor, Norman, Neil McCubbin, and John B. Sprague. 1988. "Kraft Mill Effluents in Ontario." Report to Environment Ontario. Toronto, Ontario.

Cook, C. Roger. 1988. "Organochlorine Discharges from a Bleached Kraft Mill with Oxygen Delignification and Secondary Treatment." Proceedings of the Canadian Pulp and Paper Association Environment Conference, Vancouver, BC.

Eisler, Ronald. 1986. "Dioxin Hazards to Fish, Wildlife, and Invertebrates: A Synoptic Review." U.S. Fish and Wildlife Service Biological Report 85. Laurel, MD.

Federal Register (FR). 1985. "Dioxin and Furan Pollution; Partial Grant of Environmental Defense Fund / National Wildlife Federation Citizen's Petition: 50:4426-4448.

_____. 1988. "Hazardous Waste Management System, Identification and Listing of Hazardous Waste: Use of a Generic Dilution/Attenuation Factor for Establishing Regulatory Levels and Chronic Reference Level Revisions." 53:18024-18032.

Germgard, U., and S. Larson. 1983. "Oxygen Bleaching in the Modern Softwood Kraft Mill." *Paperi ja Puu*, 4:287-290.

Grayman, Walter M. 1989. "Water Quality Simulation of Dioxin Discharges from Pulp and Paper Mills." Memorandum to the Office of Policy, Planning, and Evaluation, U.S. Environmental Protection Agency. Washington, DC.

Hall E.R., J. Fraser, S. Garden, and L.A. Cornaccio. 1988. "Organochlorine Discharges in Wastewaters from Kraft Mill Bleach Plants." *Proceedings of the Canadian Pulp and Paper Association Environment Conference*, Vancouver, BC.

ICF, Inc. December 7, 1988. "Additional TC RIA Exhibits of Risk and Resource Damage," Memorandum to the Office of Solid Waste, U.S. Environmental Protection Agency. Fairfax, VA.

_____. 1989. "Toxicity Characteristic Regulatory Impact Analysis." Draft report to the Office of Solid Waste, U.S. Environmental Protection Agency. Fairfax, VA.

McCubbin, Neil. 1988. "Calculations of Organochlorine Discharge from Pulp Mills and Estimate of Cost of Abatement Measures." Draft report to the Office of Policy, Planning, and Evaluation, U.S. Environmental Protection Agency. Washington, DC.

McLeay D. J. and Associates, Ltd. 1987. "Aquatic Toxicity of Pulp and Paper Mill Effluent: A Review." Report EPS 4/PF/1. Ottawa, Environment Canada. Ontario, Canada.

Radian Corporation. 1989. "National Air Toxics Information Clearinghouse: Natick Data Base Report on State, Local and EPA Air Toxics Activities." Report to the Office of Air Quality Planning and

Standards, U.S. Environmental Protection Agency. EPA-450/3-89-29. Austin, TX.

Sjostrom L., R. Radestrom, and K. Lindstrom. 1982. "Determination of Total Organic Chlorine in Bleach Liquors." *Svensk Papperstidn*, 85 3:7-13.

Sodergren, A., B-E Bengtsson, P. Jonsson, S. Lagergren, A. Larsson, M. Olsson, and L. Renberg. 1987. "Summary of Results from the Swedish Project Environment/Cellulose." *Proceedings Second IAWPRC Symposium on Forest Industry Wastewaters*. Tampere, Finland.

U.S. Environmental Protection Agency. 1983. "Guidelines for Regulatory Impact Analysis." EPA-230-01-89-003. Washington, DC.

_____. 1984. "Ambient Water Quality Criteria for 2,3,7,8-Tetrachloro-dibenzo-p-dioxin." EPA-440/5-84-007. Washington, DC.

_____. 1985(a). "Health Assessment Document for Polychlorinated Dibenzo-p dioxins." EPA-600-8-84-014F. Washington, DC.

_____. 1985(b). "Chloroform Exposure and Risk Assessment." Docket, Office of Air Quality Planning and Standards, Research Triangle Park, NC.

_____. 1985(c). "Health Assessment Document for Chloroform." EPA-500/8-84-004F. Research Triangle Park, NC.

_____. 1985(d). "Survey of Chloroform Emission Sources." EPA-450/3-85-026. Office of Air Quality Planning and Standards, Research Triangle Park, NC.

_____. 1986. "User's Manual for the Human Exposure Model (HEM)." EPA-450/5-86-001. Office of Air Quality Planning and Standards, Research Triangle Park, NC.

_____. 1988(a). "A Cancer Risk-Specific Dose Estimate for 2,3,7,8-TCDD." External review draft. EPA-600/6-88-007a. Washington, DC.

_____. 1988(b). "Estimating Exposures to 2,3,7,8-TCDD." External review draft. EPA 600/6-88-005a. Washington, DC.

_____. 1988(c). "Proposed Risk Management Actions for Dioxin Contamination Midland, Michigan." EPA-905/4-88-006. Chicago, IL.

_____. 1988(d). "U.S. EPA/Paper Industry Cooperative Dioxin Screening Study." EPA-440/1-88-025. Washington, DC.

_____. 1989(a). "Preliminary Data Summary for the Pulp, Paper and Paperboard Point Source Category." EPA-440/1-89-025. Washington, DC.

_____. 1989(b). "National Bioaccumulation Survey." Draft report. Washington, DC.

Westat, Inc. 1987. "Screening Survey of Industrial D Establishments." Appendix C. Draft final report to the Office of Solid Waste, U.S. Environmental Protection Agency. Rockville, MD.

Vandenberg, John. 1990. Personal communication. Office of Air Quality Planning and Standards, U.S. Environmental Protection Agency. Research Triangle Park, NC.

Voss R.H., C.E. Luthe, B.I. Flemming, R.M. Berry, and L.H. Allen. 1988. "Some New Insights into the Origins of Dioxins Formed During Chemical Pulp Bleaching." *Proceedings of the Canadian Pulp and Paper Environment Conference*, Vancouver, BC.

Appendix 8-A

MILLS IN THE TOX ANALYSIS

Table 8-A-1. Pulp Mills in the TOX Analysis ($1984 10³)

Mill & Bleach Lines[a]	Current TOX Discharge (lbs/ton)	5 lb/ton		3 lb/ton	
		Process Changes to Comply[b]	Total Annual Cost[c]	Process Changes to Comply[b]	Total Annual Cost[c]
3	7.8	1	$460	2	$890
3	7.8	1	1,010	2	1,450
3	7.8	1	1,500	2	1,910
4	4.9	0	0	1	270
5	7.8	1	390	2	800
5	3.8	0	0	1	1,400
6	7.8	1	1,700	2	1,360
8	7.8	1	490	2	650
8	4.8	0	0	1	300
9	4.9	0	0	1	170
10	7.8	1	340	2	770
10	7.8	1	410	2	770
11[d]	7.2	1	1,410	2	1,920
11[d]	7.2	1	1,800	2	2,050
13	3.5	0	0	1	300
15	7.8	1	510	2	650
15	7.8	1	1,040	2	1,230
21[d]	4.2	0	0	1	1,190
21[d]	7.8	1	1,120	2	1,420
23	7.2	1	580	2	920
24[d]	7.8	1	1,230	2	750
26	7.8	1	860	2	860
26	4.2	0	0	1	410
31	7.8	1	840	2	1,280
39	3.2	0	0	1	170
39	1.8	0	0	0	0
41	5.4	1	780	1	780
42[d]	4.2	0	0	1	150
43[d]	7.8	1	1,390	2	1,420
44[d]	7.2	1	1,030	2	1,210

(cont.)

Table 8-A-1. (cont.) Pulp Mills in the TOX Analysis ($1984 10^3)

Mill & Bleach Lines[a]	Current TOX Discharge (lbs/ton)	5 lb/ton		3 lb/ton	
		Process Changes to Comply[b]	Total Annual Cost[c]	Process Changes to Comply[b]	Total Annual Cost[c]
49	5.5	1	$30	2	$1,160
54	7.2	1	750	2	1,180
54	3.8	0	0	1	700
57	5.0	1	910	2	410
57	7.2	1	910	2	1,240
58	7.8	1	820	2	1,170
58	7.2	1	1,100	2	1,380
72	4.9	0	0	1	870
74	4.2	0	0	1	350
77[d]	6.6	1	1,200	2	1,700
80	4.0	0	0	1	120
80	7.2	1	940	2	1,000
82[d]	3.6	0	0	1	660
82[d]	7.2	1	1,090	2	1,420
89	7.2	1	910	2	1,240
93[d]	4.2	0	0	1	650
93[d]	7.8	1	730	2	1,130
94	4.2	0	0	1	270
94	7.9	1	770	2	830
99	4.2	0	0	1	100
99	7.8	1	900	2	1,330
101	4.2	0	0	1	680
103	4.8	0	0	1	460
104	4.8	0	0	1	240
105	4.8	0	0	1	820
142[d]	5.2	1	30	2	450
142	2.8	0	0	0	0
151	5.2	1	30	2	380
151	2.6	0	0	0	0
152[d]	7.8	1	1,500	2	1,930
153	0.0	0	0	0	0
155[d]	4.2	0	0	1	750
158[d]	7.8	1	1,090	2	1,090
165	4.2	0	0	1	1,050
201[d]	4.9	0	0	1	950
204	4.8	0	0	1	1,330
213[d]	7.8	1	930	2	260
213[d]	4.2	0	0	1	170
214[d]	4.9	0	0	1	600
216[d]	7.8	1	680	2	480
					(cont.)

Table 8-A-1. (cont.) Pulp Mills in the TOX Analysis ($1984 10^3)

Mill & Bleach Lines[a]	Current TOX Discharge (lbs/ton)	5 lb/ton		3 lb/ton	
		Process Changes to Comply[b]	Total Annual Cost[c]	Process Changes to Comply[b]	Total Annual Cost[c]
216[d]	3.8	0	$0	1	$110
217	2.6	0	0	0	0
217[d]	7.8	1	850	2	1,140
235	2.9	0	0	0	0
242	3.6	0	0	1	1,110
252	2.4	0	0	0	0
259	2.5	0	0	0	0
265	3.0	0	0	1	1,240
266[d]	4.9	0	0	1	1,070
272[d]	3.9	0	0	2	780
273	5.0	1	440	2	380
277	6.0	1	900	1	1,330
278	3.6	0	0	1	170
280[d]	7.8	1	1,480	2	1,910
281[d]	7.8	1	780	2	1,090
281[d]	7.8	1	500	2	890
281	2.6	0	0	0	0
282	7.8	1	1,480	2	1,910
293[d]	7.8	1	1,220	2	1,000
293[d]	3.6	0	0	1	660
296	2.1	0	0	0	0
299	5.4	1	800	1	800
301	7.8	1	850	2	1,290
301	3.0	0	0	1	620
301	4.2	0	0	1	810
303	7.2	1	1,040	2	1,390
305	7.8	1	980	2	1,280
305	7.2	1	1,080	2	1,600
306	3.0	0	0	1	820
307	3.6	0	0	1	670
310	7.2	1	1,250	2	1,720
314	4.8	0	0	1	1,430
317	3.0	0	0	1	10
318	3.0	0	0	1	830
319	2.0	0	0	0	0
329	3.0	0	0	1	1,130
331	3.0	0	0	1	1,130
341	3.0	0	0	1	160
343	2.0	0	0	0	0
352	2.5	0	0	0	0

(cont.)

Table 8-A-1. (cont.) Pulp Mills in the TOX Analysis ($1984 10^3)

Mill & Bleach Lines[a]	Current TOX Discharge (lbs/ton)	5 lb/ton Process Changes to Comply[b]	5 lb/ton Total Annual Cost[c]	3 lb/ton Process Changes to Comply[b]	3 lb/ton Total Annual Cost[c]
352	1.8	0	$0	0	$0
354	3.6	0	0	1	640
360	3.6	0	0	1	1,400
361	4.9	0	0	1	990
362	3.0	0	0	1	180
364	2.2	0	0	0	0
365	7.8	1	1,000	2	270
366	1.7	0	0	0	0
366[d]	7.8	1	810	2	380
367	7.2	1	750	2	1,170
371[d]	7.8	1	960	2	1,270
372	3.6	0	0	1	1,400
376	7.2	1	1,190	2	1,620
380	7.8	1	1,320	2	1,630
386	2.8	0	0	0	0
406	7.8	1	1,560	2	1,950
410[d]	7.8	1	730	2	1,090
410[d]	7.8	1	480	2	880
425	7.8	1	1,070	2	1,400
1441	4.2	0	0	1	720
1503	3.6	0	0	1	1,020
1503	7.8	1	120	2	490
1572	4.2	0	0	1	710
1598	7.2	1	990	2	1,370
1598	3.6	0	0	1	70
1617	4.8	0	0	1	1,120
1705[d]	4.9	0	0	1	830
Total Cost			$58,890		$111,080
Total Number of Lines/Mills Affected by Per Ton Limitations:			65/55		120/94

a. Each repetition of a mill ID number indicates an additional bleach line for that mill.
b. 0 = no changes required, 1 = high chlorine dioxide substitution,
   2 = oxygen delignification and high chlorine dioxide substitution.
c. Based on 1984 capacity.
d. Mills which would also be included under an ambient approach.

Appendix 8-B

CRITIQUE OF THE CHLOROFORM RISK ASSESSMENT

Discussion of the Key Assumptions Used in the EPA Risk Assessment

Chloroform ($CHCl_3$) is a well known chemical, which for years was used in high concentrations as a human anesthetic. At very high chronic levels (>100 ppm in drinking water), unequivocal toxic effects on rat and mice livers have been found. However, the situation in regard to carcinogenesis is not so clear cut, as there are numerous questions about the use of the linear no-threshold model, the species of animals used for the development of the risk assessment, the method by which animals were dosed, and the method by which extrapolation to humans was performed. These concerns were raised in EPA's Health Assessment Document for Chloroform [1985], and were reiterated in the critique prepared by the American Paper Institute [API. 1986].

Linear No-Threshold Model. The expression $P_t(d) = q \times d$ represents the risk assessment model used for chloroform, where $P_t(d)$ is an incremental lifetime risk of getting the cancer in question, when the individual is subjected to a lifetime dose, d, of the carcinogen in question. Clearly the model is linear and shows a finite increased risk at all doses except zero. This model appears to be valid for certain carcinogens, especially ionizing radiation. The model is predicated on the concept that a carcinogen (or one of its metabolites) reacts with genetic DNA, and in particular with that genetic portion of the DNA which disrupts cellular growth and leads to a new generation of transformed malignant cells. Chemical agents which do not directly affect genetic material, or are not shown to lead to mutational changes in cells, may act to promote or exacerbate malignant transformation, but they would not be considered to be primary carcinogens. Promoter chemicals typically would exhibit toxicity toward some cellular subsystem: for example, the DNA repair enzymes. Cytotoxic substances generally, and carcinogenic promoters specifically, usually exhibit a non-linear threshold behavior and would therefore not be candidates for risk assessment using the model shown above.

There is considerable evidence that chloroform would fall into the non-linear threshold model category. A number of animal studies have shown that, although chloroform or its metabolites bind to cellular lipids and proteins, there is little binding to genetic DNA even at cytotoxic levels. The studies of mutagenesis at the cellular level have been inconclusive with only a small percentage indicating mutagenic potential. Mutational studies of cells in whole animals have been largely negative. It should also be noted that increased liver cancers in animals occurred only in cases where chloroform levels were high enough to also cause toxic liver damage.

Reitz et al. [1990] are developing a non-linear threshold model to use for chloroform and methylene chloride risk assessment. Their model is predicated on the biological concept that chloroform induces cancer by killing cells after a threshold concentration of chloroform metabolite (i.e., phosgene) has been exceeded. After cell death, the organ tissue cells in question begin dividing at a rapid rate and the increased cell division rate magnifies the natural background propensity for cells to mutate and transform to cancer cells. This notion is supported by previous empirical studies which showed that liver tumors formed in certain animals only after significant toxicity occurred. Their work is putting the sought after pharmaco-kinetic models for carcinogenesis on a firm footing.

Concerns over the Animal Species Used. The EPA risk assessment for chloroform was based on studies using the B6C3F1 Hybrid mouse. This particular species has been shown to exhibit a very high rate of spontaneous liver tumor formation. Various experts from non-industry professional groups and foreign Government health agencies have questioned the use of this species for bioassays using liver tumors as the end point. A study using a rat species and carried out by SRI International [Jorgensen et al. 1985] was also available for the EPA risk assessment, but it was rejected in favor of the mouse study for the risk assessment. Reitz et al. have soon-to-be-published data clearly showing that chloroform metabolism in human liver tissue is similar to that of the rat and both are much lower than the mouse. Given the Reitz et al. model for carcinogenesis by chloroform, the rate of metabolism to phosgene is the critical step in causing cell death and subsequently increased division and malignancy.

Concerns About the Method of Exposure. In the National Cancer Institute [NCI. 1976] mouse study, chloroform exposure occurred through an oral gavage technique, whereby the chloroform was introduced into the animal dissolved in corn oil. In the SRI rat and mouse study, exposure occurred via drinking water, a method more similar to the human situation. The actual total dose in the SRI rat and mouse study was greater than in the NCI mouse study, but the cancer incidence was lower. On the other hand, the NCI mouse study incorporated higher dose rates for fewer months than the rat study and there is some evidence that this produced toxic effects involving a threshold above which the glutathione protective system could not cope. According to the Reitz model, this higher dose rate is a very important factor.

Furthermore, there is evidence that fats and oils can act as tumor promoters on their own. Finally, the presence of oils activates various enzyme systems, which can alter the rate of metabolism of the toxicant/carcinogen (i.e., chloroform).

It would appear that a model based on exposure via drinking water may have been a better choice for a number of reasons.

Extrapolation of Dose from Animals to Humans. In most instances, the extrapolation of risk from animals to humans is based on body weight. However, for chloroform the EPA assessment used body surface area without sufficient justification for not using body weight for extrapolation. There are arguments pro and con, but the surface area extrapolation appears to be more relevant to the case where the ingested compound is the carcinogen rather than one of its metabolites for which the body weight extrapolation may be the more relevant.

Quantitative Aspects of the Critique of the EPA Risk Assessment

If the SRI rat kidney cancer study were used instead of the NCI mouse liver cancer studies, the risk assessment would provide an upper-bound estimate of individual lifetime risk of 4.4×10^{-3} (pooled male and female mouse) versus a risk of 8.1×10^{-2} for the NCI mouse data. This provides a factor of 20 difference in risk (at a dose of one mg/kg/day).

The API [1986] critique goes further in arguing that the use of a 95 percent upper confidence level concept is inappropriate, and that the true estimate of risk should be the maximum likelihood estimate, which they claim is a more realistic estimate or actual risk. Using the maximum likelihood estimate reduces the risk by an additional factor of five.

Finally, use of a body weight rather than body surface area extrapolation would further reduce the risk by another factor of five, so that the risk differential between the EPA value and the API derived risk could be a factor of 500. Although one cannot argue the absoluteness of using one set of assumptions versus the other, it is clear that the sensitivity of risk to the choice of a few critical variables is large indeed. Furthermore, this large sensitivity exists even in the absence of any large scale changes in the kind of model (i.e., linear non-threshold) which is used to calculate risk.

References

American Paper Institute (API) and National Forest Products Association. 1986. "Comments on Hazardous Waste Management System; Identification and Listing of Hazardous Waste; Notification Requirements; Reportable Quantity Adjustments." Proposed Rule 51 Federal Regulation 21648-21693. EPA Docket F-86-TC-FFFFF.

Jorgensen, Ted A., Earl F. Murhenry, Carol Rushbrook, Richard J. Biell, and Merrel Robinson. (SRI International). 1985. "Carcinogenicity of Chloroform in Drinking Water to Male Osborne - Mended Rats and Female B6C3F1 Mice." *Fundamental and Applied Toxicology*, 5:760-769.

National Cancer Institute (NCI). 1976. "National Cancer Institute Carcinogenesis Bioassay of Chloroform." PB2640181AS. National Technical Information Service, Springfield, VA.

Reitz, R., A.L. Mendrala, R.A. Corley, J.F. Quast, M.L. Gargas, M.E. Andersen, D.A. Staats, R.B. Conolly. 1990. "Estimating the Risk of Liver Cancer Associated with Human Exposures to Chloroform." Paper submitted to the *Journal of Toxicology and Applied Pharmacology*.

U.S. Environmental Protection Agency. 1985. "Health Assessment Document for Chloroform." EPA-500/8-84-004F. Washington, DC.

Appendix 8-C

PROCESS MODIFICATION COSTS

Table 8-C-1. Process Modification Costs to Model Mills with
Existing Bleach Plant Sequences ($1984 10^3)

Existing Sequence[a]	Model Mill	New Sequence[a] / Modifications Required	EPA	Reestimated[b]
			Annual Net Costs	
C-E-H-D	Kraft - Hardwood (181 mg/day)	C-E-E_O-D Replace H stage with E_O stage. Add new oxygen storage and mixing equipment.	$168	($32)[c]
	Kraft - Softwood (181 mg/day)	C-E_O-D-D Add oxygen to extraction stage; replace H stage with D stage. New D stage equipment and chlorine dioxide (ClO_2) generator.	$1,693	$980
C-E-H-D	Kraft - Hardwood (545 mg/day)	C-E-E_O-D Replace H Stage with E stage. Add new oxygen storage and mixing equipment.	$265	($175)[c]
	Kraft - Softwood (545 mg/day)	C-E_O-D-D Add oxygen to extraction stage; replace H stage with D stage. New D stage equipment and ClO_2 generator.	$1,805	$950
C-E-H	Sulfite (181 mg/day)	C-E-D Replace H stage with D stage. New D stage equipment, ClO_2 generator, ClO_2 chemical storage tanks, handling facilities.	$1,876	$1,130

(cont.)

Table 8-C-1. (cont.) Process Modification Costs to Model Mills with Existing Bleach Plant Sequences ($1984 10^3)

Existing Sequence[a]	Model Mill	New Sequence[a] / Modifications Required	Annual Net Costs EPA	Annual Net Costs Reestimated[b]
C-E-H	Kraft (363 mg/day)	C-E$_O$-D Replace H stage with D stage. Add oxygen to extraction stage. New D stage equipment, ClO$_2$ generator, ClO$_2$ chemical storage tanks, handling facilities.	$2,197	$1,100
C-E-H-E-D	Kraft (545 mg/day)	C-E-D-E-D Replace H stage with D stage. New D stage equipment and additional ClO$_2$ generator.	$1,850	($590)[c]
C-E-H-D-E-D	Kraft (545 mg/day)	C-E-E$_O$-D-E-D Add oxygen to extraction stage. Bypass H stage. Additional ClO$_2$ generator may be required.	($25)[c]	($275)[c]

a. Sequence Key:

C =	Chlorine Stage	D =	Chlorine Dioxide Stage
E =	Caustic Extraction Stage	O =	Oxygen Stage
E$_O$ =	Oxygen added as Supplement to Caustic Extraction Stage	H =	Sodium or Calcium Hypochlorite Stage

b. The changes in the costs of chemicals and the amount of chemical usage are based on a June 30, 1983, memorandum from Rudi Schleinkofer, Ingersoll Rand to Sam Duletsky, GCA Corporation. This memorandum is available in the EPA docket. We lowered the capital recovery factor to be consistent with the analyses of other capital expenditures.

c. Net savings.

Source: "Survey of Chloroform Emission Sources" [EPA. 1985(d)].

Appendix 8-D

MILL-SPECIFIC COSTS FOR HYPOCHLORITE ELIMINATION

Table 8-D-1. Costs of Hypochlorite Elimination[a] ($1984 10^3)

Mill	Type[b]	Annual Costs	Mill	Type[b]	Annual Costs
702	BCT	$860	810[c]	MI	$200
751[c]	BCT	950	811	MI	310
754[c]	BCT	320	816	MI	450
806[d]	BCT	(920)	817[c]	MI	860
823[c]	BCT	820	835	MI	990
844	BCT	1,900	837	MI	310
847[c,d]	BCT	(700)	841[c]	MI	1,500
729[c,d]	MBK	(280)	843	MI	1,100
730[c,d]	MBK	(280)	850	MI	800
747[c,d]	MBK	(830)	854	MI	1,100
757	MBK	950	858[c]	MI	300
775[c,d]	MBK	(280)	860[d]	MI	(130)
807	MBK	450	861[c]	MI	890
867[c]	MBK	720	865[c]	MI	1,040
703[c]	MI	590	892	MI	1,090
710[c]	MI	300	805[c]	PS	1,200
719[c]	MI	1,080	830	PS	1,200
720[c,d]	MI	(390)	866[c,d]	PS	1,110
723	MI	860	869[c,d]	PS	980
728[c]	MI	850	870[c]	PS	1,100
732[d]	MI	(280)	871	PS	1,130
735[d]	MI	(90)	874	PS	1,130
736[c]	MI	970	876[c]	PS	1,130
739[c]	MI	990	877	PS	1,130
759	MI	1,600	879[c]	PS	450
768[c,d]	MI	(220)	883	PS	500
781[d]	MI	(390)	887	PS	1,090
788	MI	590	707	SK	850
795[d]	MI	(390)	764	SK	860
800	MI	1,030	771	SK	1,000

(cont.)

Table 8-D-1.   (cont.) Costs of Hypochlorite Elimination[a] ($1984 10^3)

Mill	Type[b]	Annual Costs	Mill	Type[b]	Annual Costs
774[c]	SK	$910	828	SK	$1,090
782[c]	SK	910	829[d]	SK	(180)
785	SK	1,030	831[c]	SK	950
791	SK	920	842	SK	1,100
813[c,d]	SK	(180)	845	SK	950

a. EPA estimates.
b. Abbreviations:

BCT	=	Best Conventional Technology
MBK	=	Market Bleached Kraft
MI	=	Miscellaneous Integrated
PS	=	Papergrade Sulfite
SK	=	Soda and Kraft

c. Mills which would be included under an ambient approach.
d. Mills which would be included under a benefits approach.

Appendix 8-E

MILL-SPECIFIC COSTS FOR CHLOROFORM REDUCTION

Table 8-E-1.   Annual Costs Used in Evaluating Alternative Regulatory
Approaches for Chloroform as a Hazardous Waste
($1984 10^3)

Mill	Type[a]	Annual Cost	Mill	Type[a]	Annual Cost
3	BK	$1,470	204[b,c]	PS	$30
5	BK	830	213[b]	BK	1,460
8	AF	1,120	214[b]	BK	1,380
9	BK	900	235	AF	1,360
13	BK	10	239	DI	690
23	BK	780	248	BK	650
26	AF	1,440	250[b,c]	PS	30
31[b]	AF	420	252	AF	190
49	BK	2,010	259	AF	460
54	BK	1,400	260	DI	140
58	BK	10	265[c]	PS	25
74	BK	90	266[b]	AF	1,260
80	BK	1,180	281	BK	1,790
82	AF	210	292	DI	30
93	AF	1,230	314[c]	PS	30
94	AF	10	318[c]	PS	30
96	DI	420	321	DI	1,000
99	AF	1,390	325[b]	DI	340
103	BK	150	327	DI	80
142	AF	1,310	329[c]	PS	20
152	AF	1,380	335[c]	PS	50
164	DI	60	343[c]	PS	20
183	AF	690	344	DI	210
200[b]	DI	1,220	347	DI	120

a. Abbreviations: BK = Bleached Kraft
AF = Alkaline Fine
PS = Papergrade Sulfite
DI = De-inking
b. Mills which would be included under an ambient approach.
c. Mills which would be included under a benefits approach.

Chapter 9

CONCLUSIONS

The results of our analyses may come as a surprise to the various schools of environmental management in the United States. The utilitarian school, which reflects the concern of the public health profession, supports the use of ambient-based standards as a scientific and reasonable way to protect health and welfare [Rodgers. 1986]. The absolutist school, which reflects the interests of the environmental rights-oriented movement, supports the use of technology-based standards as a necessary and practical approach [Latin. 1985]. The rationalist school, which reflects the economist's view of environmental management, supports the use of benefits-based standards as a balanced and sensible approach [Krier. 1974, Freeman. 1980, Pedersen. 1988].

The utilitarian school dominated environmental management until the passage of the Clean Water Act in 1972 and has been in retreat since then. The absolutist school has dominated environmental management since the Clean Water Act of 1972, as reflected recently in the Resource Conservation and Recovery Act of 1984 and the Safe Drinking Water Act of 1987. The rationalist school has prevailed only with chemical regulating statutes because the Congress recognized the importance as well as the hazards of chemicals to society. The few rationalist concessions in other pollution reduction statutes, such as odor control and lead in gasoline, appear to be more an oversight than a deliberate decision.

This chapter first makes some general observations about the relative effectiveness of the three regulatory approaches, then examines Congressional motives for adopting the approaches it did, and finally offers some recommendations for environmental policy in the future.

REGULATORY APPROACHES

The lessons to be learned from our historical analyses (1973-1984) differ from those gained from our analyses of future projections (1984-

1994). This is because the application of ambient-based standards has often been tempered by cost-effectiveness considerations. These adjustments to meet individual circumstances are not taken into account in projections of future performance.

Historical Analysis (1973-1984)

Our historical analysis reflects the results of implementing the regulatory approaches rather than preconceived ideas of how they would work. As seen in Table 9-1, the technology-based standards for water pollution management failed as a strategy for generating net benefits. The costs clearly exceeded the benefits in the aggregate, as well as in the specific in most situations. Benefits exceeded costs at only seven of the 68 mills investigated in our study. The ambient-based standards for ambient air pollution management succeeded as a strategy for generating net benefits. The benefits exceeded the costs in the aggregate as well as in the specific for about one-third of the mills (20 of 60 mills). We did not expect emission reductions at all mills to exhibit benefits greater than costs because the ambient standard was uniformly implemented across the country. The benefits-based standards for air pollution management also

Table 9-1. Benefits and Costs Attributable to the Three Approaches -- 1973-1984[a] ($1984 10^6)

Approach	Mills with B>C / Total Mills	Total Benefits[b]	Total Costs	Net Benefits
Technology (CWA)	7 / 68	$36.6	$96.6	($60.0)
Ambient (CAA)	20 / 60	$25.1	$24.2	$0.9
Benefits (CAA)	29 / 60	$86.9	$55.8	$31.1

a. Data taken from Chapters 3, 4, and 5.
b. Total benefit estimates based on mid-point calculations. For total water benefits, the mid-point is the average of the minimum and maximum values.

succeeded as a strategy for generating net benefits. The benefits exceeded the costs in the aggregate as well as in the specific for about one-half of the mills. Benefits exceeded costs at 29 of the 60 mills investigated. The benefits-based standard failed in many specific situations, partly because the EPA guidance in 1979 did not require rigorous quantitative analysis of both benefits and costs.

These analyses of the failures of the technology approach call into question the wisdom of the absolutist approach to environmental management. Not only did the technology-based standard for water pollution fail to produce net benefits but it also failed on several other counts that should be important to the absolutist school. Approximately one-quarter of the mill environments investigated still remain degraded after implementation of the technology-based standard. Focusing on effluent limits as the measure of success apparently restricts thinking about the real aim of environmental management, which is protection of human health and welfare. Similarly, the compliance investigation (Chapter 6) found many more mills violating ambient water quality standards than ambient air quality standards. Even recognizing the limitations of the EPA data, the fact that 30 mills potentially are not in compliance with the ambient water quality standards, as compared with seven mills potentially not in compliance with the ambient air quality standards, points to serious deficiencies in the technology approach for correcting environmental problems.

The relative effectiveness of the ambient approach calls into question the harshness of the rationalist school's critique of the utilitarian school. The rationalist school rejects not only the technology approach but also the ambient approach because it fails to incorporate population exposure and the costs of meeting a given standard.

This preconceived failure did not happen in the implementation of the ambient-based TSP standard around pulp and paper mills, however. In fact, the ambient-based standard produced results that were remarkably similar to the benefits-based standard for TRS in the aggregate and in the specific. The ambient-based standard succeeded as well as it did, we think, because local environmental managers did not require reductions in TSP emissions from pulp-producing mills where the costs per ton of emission reduction were high and the potential human health risks were small. The Clean Air Act focused the attention of these environmental managers on achieving area-wide ambient air quality, and as a result allowed them to take into account the cost-effectiveness of emission reductions among different sources (primarily manufacturing and utilities). Our investigation found considerable variation in the cost per ton of TSP reduction at pulp-producing mills in different areas. Other investigations have found considerable variation among different source categories [Crandall. 1983].

To our knowledge, no one has attempted to relate these variations in specific areas to the relative cost effectiveness of emission reduction and population exposure.

Projections of the Future (1984-1994)

The results of our analyses of alternative regulatory approaches in the future are best summarized separately for conventional and toxic pollutants. The analyses of conventional pollutants, covering both compliance and growth, generally make a strong argument for the benefits approach. As seen in Table 9-2, the benefits approach for compliance is the only approach that would produce net benefits. The technology approach for water compliance, but not air compliance, might also generate net benefits, however. The technology approach for water compliance, in contrast to air compliance, would result in many mill-

Table 9-2.   Benefits and Costs Attributable to the Three
Compliance Strategies ($1984 10^6)

Approach	Mills with B>C / Total Mills	Total Benefits[b]	Total Costs	Net Benefits
		---- Water ----		
Technology	6 / 49	$16.5	$19.1	($2.6)
Ambient	5 / 9	$ 4.9	$ 9.7	($4.8)
Benefits	10 / 10	$13.5	$ 1.8	$11.7
		---- Air[c] ----		
Technology	1 / 7	$ 0.4	$ 2.7	($2.3)
Ambient	1 / 7	$ 4.6	$ 6.9	($2.3)
Benefits	3 / 3	$ 4.2	$ 2.0	$2.2

a. Data taken from Chapter 6.
b. Total benefit estimates based on mid-point calculations. For total water
   benefits, the mid-point is the average of the minimum and maximum values.
c. TSP only.

specific, positive benefit-cost outcomes. This is because the existing technology-based standards have constrained pollutant reductions needed to meet ambient water quality standards.

Our analysis of the relative efficienty of the three regulatory approaches for reducing the adverse pollution effects resulting from industrial growth is summarized in Table 9-3. This shows that only the benefits approach for projected water (BOD) and air (TSP) problems

Table 9-3. Benefits and Costs Attributable to the Three Approaches to Industrial Growth[a] ($1984 10^6)

Approach	Mills with B>C / Total Mills	Total Benefits[b]	Total Costs	Net Benefits
		---- Water ----		
Technology	28 / 176	$33.5	$101.1	($67.6)
Ambient	3 / 49	$ 4.8	$161.0	($156.2)
Benefits	16 / 16	$21.4	$4.3	$17.1
		---- Air[d] ----		
Technology	7 / 59	$10.0	$49.1	($39.1)
Ambient	1 / 33	$ 1.3	$19.4	($18.1)
Benefits	5 / 5	$ 6.7	$ 3.6	$ 3.1
		---- Air[e] ----		
Technology	11 / 32	$ 9.3	$ 9.3	($ 0.0)
Ambient	9 / 25	$ 7.4	$ 7.3	($ 0.1)
Benefits	11 / 11	$ 8.1	$ 3.2	$ 4.9

a. Data taken from Chapter 7.
b. Total benefit estimates based on mid-point calculations. For total water benefits, the mid-point is the average of the minimum and maximum values.
c. Total costs reflect only those mills subject to water or air quality modeling.
d. TSP only.
e. TRS only.

would generate net benefits. All three approaches would generate net benefits for projected air (TRS) problems. Even for projected air (TRS) problems, however, the benefits approach is preferable on the basis of generating net benefits. Neither the technology-forcing nor the antideterioration approaches would result in efficient outcomes except in the case of TRS.

These findings about the ambient approach deserve special attention because they contradict the findings about the efficiency of the ambient approach in the past. We speculate that a future-oriented ambient approach is not efficient because our evaluation of the future-oriented approaches does not consider how local environmental managers, particularly those dealing with air pollution problems, will make allowances for variations in cost-effectiveness and population exposures. This evaluation points to the limits of critiquing regulatory approaches only from an ex-ante perspective, as is most often done by the rationalist school, and strongly supports the necessity for more rigorous post-evaluations of alternative regulatory approaches. In addition, we recognize that ambient-based standards for all pollutants may not produce net benefits. According to the U.S. Office of Technology Assessment [1989], partial compliance with the ozone ambient standard by the year 2004 would not generate net benefits.[1]

Limitations in data and analytical techniques, as well as the resources available to evaluate the environmental risks from the pollutants of concern, constrained our analyses of toxic pollutants. Our findings suggest that regulating some toxic pollutants -- dioxin as a toxic water pollutant and chloroform as a hazardous air pollutant -- would generate net benefits. While only a benefits approach would generate net benefits for chloroform as a hazardous air pollutant, any of the three approaches for dioxin in wastewater would be efficient. Our findings also suggest that reducing the other toxic pollutant -- chloroform as a hazardous waste constituent -- would not result in net benefits under any approach. The costs of regulating chloroform as a hazardous waste appear to be disproportionate to the reduction in health risks. If EPA were to regulate chloroform as a hazardous air pollutant, however, most of the chloroform in pulp and paper mill wastes would disappear.

These findings about the efficiency of regulating specific toxic pollutants may be less important than demonstrating that we can analyze the necessity for and consequences of regulating them. The absolutist school of environmental management defends its advocacy of the technology approach for toxic pollutants primarily on the grounds that environmental managers cannot evaluate the benefits from reducing toxic pollutants. Although there are uncertainties in our analyses of the benefits from alternative approaches, these analyses do show that some toxic pollutant problems entail more health effects risks than others and that some alternative approaches for reducing these are more efficient than others.

HISTORICAL PERSPECTIVE

The results of our historical analyses caused us to ask why Congress in 1972 adopted a technology approach for improving water quality when it had, as recently as 1970, employed an ambient approach for improving air quality. Admittedly, no one in the early 1970s had completed a comprehensive post-evaluation of technology versus ambient standards. However, several opponents of technology-based standards questioned them based on theoretical grounds or geographically specific case studies.

Although we could not find a convincing scholarly explanation of the shift from ambient to technology standards, we speculate that historical precedent and executive-legislative branch competition influenced the change in Congressional thinking. First, the newly created EPA in its attempt to deal with the limitations of the Clean Water Act of 1965 resurrected the 1899 Refuse Act. EPA used the Refuse Act to justify its requirements that each discharge source obtain a permit and that it meet source-specific effluent limitations. Even before Congress had passed the Clean Water Act of 1972, EPA had prepared provisional technology-based standards for most industrial dischargers. Whereas Congress did not have such a technology-based program to approve when it amended the Clean Air Act in 1970, it did have a working example to incorporate into the Clean Water Act.

Second, in the early 1970s the executive branch in the person of President Nixon and the legislative branch in the person of Senator Muskie were competing with each other to be "Mr. Clean" [Elliott, Ackerman, and Millian. 1985]. They initially sparred over the 1970 amendments to the Clean Air Act, but the final struggle, preceding the 1972 presidential election, was over the 1972 amendments to the Clean Water Act. Apparently, they kept upping the ante with more stringent provisions without the advice of the yet unfounded national environmental groups and of unorganized industrial groups. The outcome was a technology-based effluent discharge program that virtually obliterated the ambient mandate in the earlier statutes.

Our findings also caused us to ask why Congress has not passed pollution control statutes that call for more benefits-based standards. Congress has only mandated benefits-based standards for regulation of chemicals used by society. The risk-benefit provisions of the Toxic Substance Control Act and the Federal Insecticide, Fungicide, and Rodenticide Act show that Congress is concerned with achieving a reasonable balance between the benefits and risks to society when the risks are believed to be inherent in the products themselves. Apparently Congress doesn't want to acknowledge, except in some specific provisions of the three pollutant-reduction statutes, that the manufacture of desirable products, such as paper, contain desirable as well as undesirable attributes and that elimination of these undesirable attributes in the extreme can be very costly in relationship to the environmental protection achieved. An extreme outcome of Congressional directives is the previously mentioned hazardous air pollutant mandate of the Clean Air Act. Recently, it has required EPA to issue regulations that would reduce risks at a cost of over $150 million per cancer incidence [FR. 1990].

Supporters of the rationalist school of environmental management should not be in complete despair, however. They should remember that Congress mandated water development projects for over 100 years before it institutionalized in 1936 the requirement that the benefits should exceed the costs of such projects [Kneese. 1985]. The bureaucracy then took until 1951 to issue the first guidance (Green Book) on this subject. The latest guidance was issued in 1983 [U.S. Water Resources Council. 1983]. Even this guidance requires that economic efficiency be only one of three criteria for judging the merits of a water development project. Also, supporters of the rationalist school should recognize that benefit-cost evaluations are incorporated to some degree in major EPA decisions because of a series of executive orders that date back to the Nixon administration [EPA. 1987(a)]. Executive Order 12291 (1981), the current executive order, calls for an explicit benefit-cost analysis of major regulations. Unfortunately, Executive Order 12291 has generated considerable controversy about the legality and wisdom of the Office of Management and Budget's effort to oversee and coordinate the regulatory process [Morrison. 1986, Steinberg. 1986].

FUTURE DIRECTION

How can EPA focus more of the nation's resources as well as its own administrative resources on implementing regulations that generate net benefits? Certainly, EPA's implementation of technology-based standards has not eliminated significant risks and has resulted in some needless investments by public and private enterprises. Also, EPA's use of ambient-based standards has worked only when pollutant discharge is great and local environmental managers temper, for better or worse, the actual implementation of ambient standards. Lastly, EPA uses benefits-based standards infrequently, only when required by statute, because its organizational culture is dominated by meeting statutory requirements rather than by achieving measurable environmental results.

In reality, EPA can do little, given its current statutory authority. Congress must give EPA new direction, if EPA and its state counterparts are to protect health and welfare. Congress should require that EPA have a focused mission statement that encompasses its major statutory authorities. EPA's current mission statement -- reduction of risks to human health and the environment -- does not set any priority among the multitude of environmental risks that exist in our society. Instead, it justifies working on reducing any risks, significant or insignificant. At the danger of sounding simplistic, we believe that a slight modification, although one that would require statutory approval and have revolutionary implications, would fundamentally change how EPA operates. The mission of EPA should be modified to read: "reduction of *unreasonable* risks to human health and the environment." Reduction of unreasonable risks, as defined by the Conservation Foundation [1988], would focus EPA actions on those regulations and programs where the benefits to society would be comparable to the costs imposed on individuals and public and private enterprises.

Congress should also give EPA more flexibility in implementing its numerous statutes. It should require EPA to prepare and systematically revise a strategic plan that would indicate EPA's view of the most significant unreasonable risks confronting American society.[2] Congress should then allow EPA to implement individual statutory provisions consistent with its strategic plan. In so doing, Congress would free EPA from arbitrary timetables and individual statutory provisions that are of secondary importance for protecting human health and welfare.

If these two recommendations are too radical, Congress could make a modest reform in those EPA statutes that require technology-based standards. As documented in several places in this book, technology-based standards fail as a generic regulatory approach to reduce environmental risks because they do not take into account the impacts of pollutant removal on ambient quality or the size of the population that would benefit from an improvement in ambient quality. They also do not take into account the costs of pollutant reduction. Congress could require that EPA incorporate all of these factors in its decisions about the degree of stringency of technology-based standards. Although this requirement would not mandate that there should be a reasonable balance between benefits and costs, it would temper EPA's tendency to pursue technology for technology's sake alone. (See Ackerman and Stewart [1985] and Haigh et al. [1984] for similar proposals.)

Nowhere is this requirement more necessary than for hazardous waste management [EPA. 1987(b)]. EPA's recently proposed standard for municipal solid waste management would impose requirements so stringent that, because of the few people exposed, the cost per cancer incidence avoided would be $30 billion [Temple, Barker, and Sloane Inc. 1987]. Similarly, many of EPA's Superfund cleanups cost millions of dollars, yet are estimated to result in only minimal improvement in the quality of groundwater sources that no one will use for the foreseeable future. Hazardous waste management has only recently become a focus for environmental regulation. Hence, it provides an excellent opportunity to apply the lessons we have learned about the necessity of balancing benefits and costs when setting environmental standards.

Although administering benefits-based standards requires more complexity than administering uniform technology-based standards, the administrative tasks are no more complex than those already undertaken for ambient-based standards. Balancing benefits and costs will require difficult and sometimes uncertain analysis, but EPA's evaluations of the current ambient-based standards indicate that much of this analysis is being taken into consideration already, at least implicitly [EPA. 1987(a)]. Thus, applying the benefits approach may not be such a radical departure from current practices. The change will be more in the decision criterion used (net benefits vs. ambient quality), than in the way things are done. As our analysis shows, simply changing the decision criterion could lead to a vast improvement in the net benefits that society realizes from environmental management.

Notes

1. The U.S. Office of Technology Assessment (OTA) [1989] assessed to the extent possible the benefits and costs of attaining the current ozone standard by the year 2004. OTA estimated the annual costs of a 33 percent reduction in volatile organic compounds (VOC), the major precursor of ozone, at \$4.6 billion in 1994 and \$8.4 billion in 2004. They were unable to estimate the total cost of attaining the ozone standard because they could not identify (and cost) additional VOC control measures needed to meet the standard. Their mid-range estimates of annual benefits for a 35 percent reduction in VOC were \$0.2 to \$1.4 billion. The total range of estimates for a 35 percent reduction were \$0.1 to \$3.4 billion.

2. EPA recently instituted a strategic planning process [EPA. 1989]. The stated intention of the process is to promote and facilitate a more rational allocation of the limited resources available to the agency among its activities to reduce risks to health and the environment. The process will be difficult because EPA does not have a focused mission statement, the first step in a strategic planning process.

References

Ackerman, Bruce A., and Richard B. Stewart. 1985. "Reforming Environmental Law." *Standford Law Review*, 37:1333-1365.

Crandall, Robert W. 1983. *Controlling Industrial Pollution: The Economics and Politics of the Clean Air Act*. The Brookings Institution, Washington, DC.

The Conservation Foundation. 1988. "The Environmental Protection Act." Second draft. Washington, DC.

Elliott, E. Donald, Bruce A. Ackerman, and John C. Millian. 1985. "Toward a Theory of Statutory Evolution: The Federalization of Environmental Law." *Journal of Law, Economics and Organization*, 1:313-340.

Federal Register (FR). 1990. "National Emission Standards for Hazardous Air Pollutants: Benzene Emissions from Chemical Manufacturing Process Vents, Industrial Solvent Use, Benzene Waste Operations,

Benzene Transfer Operations, and Gasoline Marketing System." 55:8292-8295. Washington, DC.

Freeman, A. Myrick III. 1980. "Technology-Based Effluent Standards: The U.S. Case." *Water Resources Research*, 16:21-27.

Haigh, John A., David Harrison Jr., and Albert L. Nichols. 1984. "Benefit-Cost Analysis of Environmental Regulation: Case Studies of Hazardous Air Pollutants." *Harvard Environmental Law Review*, 8:395-434.

Kneese, Allan. 1985. *Measuring the Benefits of Clean Air and Water.* Resources for the Future, Washington, DC.

Krier, James E. 1974. "The Irrational National Air Quality Standards: Macro and Micro Mistakes." *UCLA Law Review*, 22:323-342.

Latin, Howard. 1985. "Ideal Versus Real Regulatory Efficiency: Implementation of Uniform Standards and 'Fine-Tuning' Regulatory Reforms." *Stanford Law Review*, 37:1267-1332.

Morrison, Alan B. 1986. "OMB Interference with Agency Rulemaking: The Wrong Way to Write a Regulation." *Harvard Law Review*, 99:1059-1074.

Pedersen, William F. 1988. "Turning the Tide on Water Quality." *Ecology Law Quarterly*, 15:69-102.

Rodgers, William H., Jr. 1986. *Environmental Law: Air and Water.* West Publishing, St. Paul, MN.

Steinberg, Robert E. "OMB Review of Environmental Regulations: Limitations on the Courts and Congress." *Yale Law and Policy Review*, 4:404-425.

Temple, Barker & Sloane, Inc. December 11, 1987. "Draft Regulatory Impact Analysis of Proposed Revisions to Subtitle D Criteria for Municipal Solid Waste Landfills." Report to the Economic Analysis Staff, Office of Solid Waste. U.S. Environmental Protection Agency, Washington, DC.

U.S. Environmental Protection Agency (EPA). 1987(a). "EPA's Use of Benefit-Cost Analysis: 1981-1986." EPA-230-05-87-028. Washington, DC.

_____. 1987(b). "Unfinished Business: A Comparative Assessment of Environmental Problems." Office of Policy, Planning, and Evaluation, Washington, DC.

_____. 1989. "EPA Strategic Planning and Budgeting." Memorandum of Administrator Reilly. Washington, DC.

U.S. Office of Technology Assessment. 1989. *Catching Our Breath: Next Steps for Reducing Urban Ozone.* OTA-0-412. U.S. Government Printing Office. Washington, DC.

U.S. Water Resources Council. 1983. *Economic Environmental Principles and Guidelines for Water Related Resources: Implementation Studies.* Washington, DC.

Appendix A

THE PULP AND PAPER INDUSTRY

The pulp and paper industry was selected as the focus of this study not only because it is one of the nation's largest industries but also because it is one of the largest industrial sources of pollutants. With sales of over $60 billion and a workforce of over 600,000 in 1982, pulp and paper ranked as the nation's ninth largest manufacturing industry [PPI.1985]. The industry is the largest industrial user of water and is a significant source of air pollutants.

For the purposes of this study, we identified a total of 552 operating mills manufacturing pulp, paper, and paperboard products in 1984. Of these mills, 334 discharged effluents directly to a water body. The analyses of water pollution regulations included in this book focus on 306 of the 334 direct-discharging mills. These mills accounted for 85 percent of the paper and paperboard production in 1984.The remaining mills discharged to a municipal or jointly owned wastewater treatment system. When examining the air pollution potential of the 552 operating mills, we identified 176 pulp-producing mills as the major sources of air emissions for the pulp and paper industry. The analyses of air pollution regulations focus on 157 of the 176 mills, which accounted for 95 percent of the pulp production in 1984.

This appendix provides a brief description of the industry and, more importantly, an explanation of the industry's standard manufacturing processes and the kinds of environmental pollutants commonly associated with these processes. This is intended to provide a background for those readers unfamiliar with the pulp and paper industry.

INDUSTRY PROFILE

Although the pulp and paper industry is typified by large, integrated companies, there are still many paper mills that are operated by small, specialty producers. The largest companies, such as Georgia-Pacific and

Weyerhaeuser, had sales in 1984 in excess of $5 billion [PPI. 1985]. These companies are totally integrated: growing a significant portion of their own timber, making their own pulp, and producing finished paper and paperboard products. In 1982, the 15 largest U.S. and Canadian companies accounted for 57 percent of the world pulp capacity. Whereas the largest paper mills in the United States produce over 2,000 tons of paper each day, the smallest produce less than 10 tons [Brandes. 1984]. Small companies, such as Windsor Stevens in Connecticut, Brown Products in New Hampshire, and McIntyre Paper in New York, purchase pulp from the larger companies and produce paper, generally for specialized markets.

Capacity

Table A-1 lists total annual pulp capacity by pulp type for 1973, 1984, and 1994.[1] The most notable changes between 1973 and 1984 were the increased share of sulfate pulp of total pulp capacity from 68 percent to 72 percent, and the increased share of bleached and semi-bleached of total pulp capacity, from 31 percent to 38 percent.

Table A-1 also presents production statistics grouped under the major classifications of paper, paperboard, and construction products. The most notable changes between 1973 and 1984 were the increased share of paper, from 44 percent to 49 percent of total capacity, and the increased share of printing and writing products, from 22 percent to 26 percent of total capacity.

Geographic Distribution

The 550 operating pulp and paper mills in 1984 were distributed in four major geographic regions -- the Northeast, the Midwest, the South, and the West (Figure A-1). By segmenting the industry into four regions and into its three traditional markets -- pulp, paper, and paperboard -- we can identify distinct regional trends in the U.S. pulp and paper industry.

The oldest important producing region is the Northeast, which consists of New England and the Mid-Atlantic states of New York, New Jersey, and Pennsylvania. Production next appeared in the Midwest states. These two regions together accounted for 16 percent of domestic pulp production in 1984 and for 34 percent of paper and paperboard production [API. 1985].

Table A-1. Pulp, Paper, and Paperboard Capacity --
1973, 1984, and 1994

	Capacity (10^6 short tons)		
	1973[a]	1984[b]	1994[c]
Pulp Type			
Dissolving	1.8	1.5	1.7
Chemical Paper Grades	36.6	45.5	51.7
Sulfite Paper Grades	2.4	1.7	1.9
Sulfate & Soda Paper Grades	34.3	43.7	49.6
Bleached & Semibleached Softwood	8.8	12.5	14.2
Bleached & Semibleached Hardwood	6.9	9.9	11.2
Unbleached	18.5	21.4	24.3
Semichemical	4.2	4.6	5.2
Mechanical	5.8	6.1	6.9
Screenings	0.1	0.0	0.0
Wood Pulp for Construction Paper and Board	2.2	1.3	1.5
TOTAL	50.7	59.0	67.0
Product			
Paper	27.6	37.1	41.0
Newsprint	3.6	5.9	6.5
Printing Writing and Related	13.6	19.9	22.5
Packaging and Industrial Converting	6.1	6.1	6.0
Tissue	4.3	5.3	6.0
Paperboard	30.5	36.0	42.5
Construction Products	3.9	2.8	2.5
Total	62.0	75.9	86.0

a. API. [1974].
b. API. [1985].
c. Estimated for this study. The basis for the paper and paperboard estimate is explained in Appendix 3-B. Pulp estimate for 1994 is the projected paper and paperboard capacity ratio of 1984 pulp to paper and paperboard capacity.

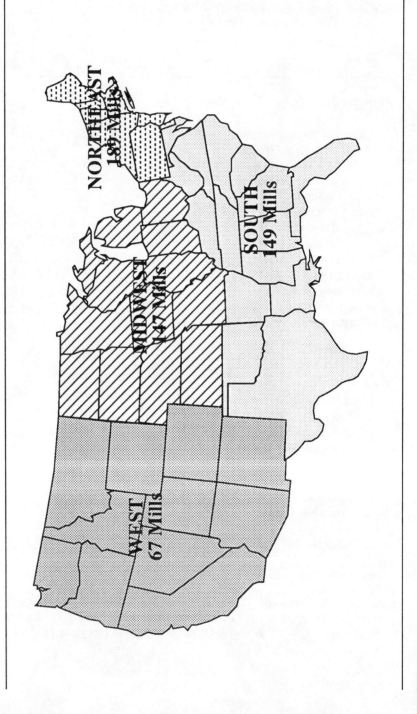

Figure A-1. Location of Pulp and Paper Mills Operating in 1984

The Northeast and Midwest regions also contain 24 percent of the commercial forests in the United States. These forests, however, are primarily hardwood, which is not suitable for many of today's high-volume pulp and paper-making technologies. Also, ownership consists in most cases of small holdings, which do not lend themselves to intensive wood production. For these reasons, the share of total wood pulp production in these regions has declined from 1973.

The relative decline of pulp production in the Northeast and Midwest states has affected the geographic distribution of paper and paperboard production as well. As integration has become economically more attractive, the production of high-volume, nonspecialized paperboard has followed the geographic shift of wood pulp production.

The South has developed into the major producer of wood pulp, paper, and paperboard in the United States. Consisting of the 14 states ranging from Maryland to Florida and westward to Texas, the Southern region possesses 45 percent of the nation's primary forest reserves, including abundant fast-growing softwoods. The topography and climate make these reserves well suited to easier harvesting and intensive management. The South produces more pulp than paper and paperboard, accounting for a large portion of U.S. market pulp. The region's share of total U.S. pulp production has been increasing regularly since 1950 and is likely to continue to grow. In 1984, the region accounted for 67 percent of all U.S. pulp production and 51 percent of total national production of paper and paperboard.

The Western states, including both Pacific and Rocky Mountain states, account for the remainder of U.S. production of pulp, paper, and paperboard. The region's softwood forests supported the production of 17 percent of the wood pulp manufactured in the United States in 1984. The West produces slightly less paper and paperboard than pulp, accounting for 14 percent of the nation's total in 1984. Forests in the Western part of the United States represent a sizable reserve of wood fiber that is being retained primarily for saw timber. For the most part these forests have not yet been tapped for the production of pulp. The terrain is mountainous and the area is isolated from most major markets (with the important exception of the export markets in the Far East).

Just as there are significant differences in the geographic distribution of wood reserves, there are also important regional differences in the types of wood in these reserves. This, in turn, results in regional differences in the distribution of production by product and process type. The South and

Pacific Northwest produce virtually all kraft linerboard, unbleached paper, and bag paper in the United States using softwood fiber. The older mills of the Northeast and Midwest produce primarily the coated and fine paper used for business papers, printing, and writing papers. Much tissue paper is also made in the Northeast and Midwest, as they are the primary markets for the product, and mixed hard and softwood forests are appropriate for these products. Building paper and board products are usually manufactured near urban areas because of their high use of fiber from waste paper and because those areas are the major markets for those products.

STANDARD MANUFACTURING PROCESSES

The production of pulp, paper, and paperboard involves several standard manufacturing processes, including: (1) preparing raw material, (2) pulping, (3) bleaching, (4) papermaking, and (5) converting [EPA. 1982]. These are diagrammed in Figure A-2.

Preparing Raw Material

Depending on the form in which the raw materials arrive at the mill, log washing, bark removal, and chipping are steps in preparing wood for pulping. Some of these processes can require large volumes of water, but the use of dry bark removal techniques or the recycling of wash water or water used in wet barking operations significantly reduces water intake.

Virgin wood is the most widely used fiber source in the pulp, paper, and paperboard industry. However, over the period covered by this study, secondary fiber sources, such as waste paper of various classifications, have gained an increasing share of raw material inputs [API. 1974, 1987]. In the early 1970s, secondary fiber sources constituted 21 percent of the raw material used in paper and paperboard manufacture. By the late 1980s, it constituted 24 percent of the raw material [API. 1974, 1987].

Pulping

Pulping is the modification of a cellulosic raw material into a form suitable for processing into paper or paperboard. Pulping processes vary from simple mechanical action, as in groundwood pulping, to complex chemical processing, such as the sulfate and sulfite processes.

Figure A-2. Pulping and Papermaking Processes and Sources of Wastes

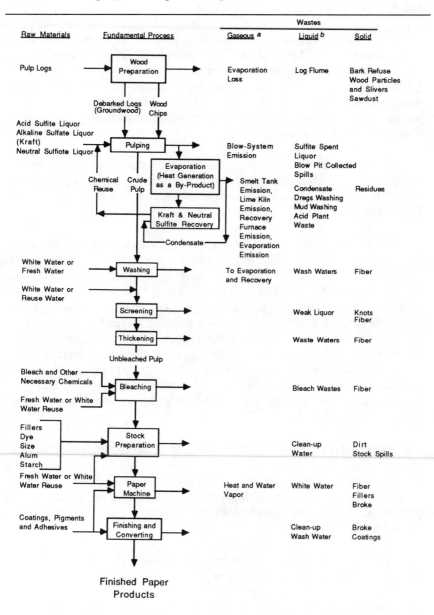

Semichemical pulping is an intermediate step between groundwood pulping and full chemical pulping.

Mechanical Pulping.   Mechanical pulp is commonly known as groundwood pulp. There are two basic processes: (1) stone groundwood, where pulp is made by tearing fiber from the side of short logs using grindstones, and (2) refiner groundwood, where pulp is produced by passing wood chips through disc refiners.   Mechanical pulping is not meant to separate the cellulose fiber in wood from the lignin, as is the objective of full chemical pulping.

In the chemi-mechanical modification of the groundwood process, wood is softened with chemicals to reduce the energy required for grinding. In a relatively new process called thermo-mechanical pulping, chips are first softened with heat and then disc-refined under pressure.

Mechanical pulps are characterized by yields of over 90 percent of the original substrate.   The pulp produced is relatively inexpensive and requires minimal use of forest resources because of these high yields. Because mechanical pulping processes do not remove the natural wood binders (lignin) and resins inherent in the wood, mechanical pulp deteriorates quite rapidly.   The pulp is suitable for use in a wide variety of consumer products, including newspapers, tissue, catalogs, one-time publications, and throw-away molded items.   Natural oxidation of the impure cellulose causes an observable yellowing early in the life of such papers. Also, a physical weakening soon occurs.   Thus, groundwood pulp must be combined with chemical pulps in order to manufacture higher-quality grades of paper.

Chemical Pulping.   Chemical pulping involves the use of cooking chemicals to achieve separation of cellulose from lignin to yield a variety of pulps with unique properties.   Chemical pulps are converted into paper products that have relatively higher quality standards or require special properties.   Three basic types of chemical pulping are now commonly used: (1) kraft, (2) sulfite, and (3) semichemical.

Kraft.   The first alkaline pulping process (developed in   the nineteenth century) was the soda process.   This was the forebear of the sulfate process, commonly referred to as the kraft process.   This process involves an alkaline treatment with   solutions of sodium sulfide and sodium hydroxide.

Several major process modifications and achievements have resulted in the widespread application of the kraft process. First, because of the increasing costs of chemicals and energy, chemicals must be recovered for economic reasons. In the 1930s, successful recovery techniques were applied to this process; these techniques have vastly improved in recent years. Second, the process was found to be adaptable to nearly all wood species. Its application to the pulping of southern pines resulted in a rapid expansion of kraft pulping in the South. Third, new developments in bleaching of kraft pulps (primarily the use of chlorine dioxide) spurred another dramatic growth period in the late 1940s and early 1950s. Use of this bleaching agent enables production of high-brightness kraft pulps that retain strength. Fourth, the development of various paper coatings has expanded the range of products that can use kraft pulps.

Early in the twentieth century, the kraft process became the major competitor of the sulfite process for some grades of pulp. Kraft pulp now accounts for about 95 percent of chemical pulp production [API. 1987].

Sulfite. Initially, sulfite pulping involved the use of calcium (lime slurries sulfated with sulfur dioxide) as the sulfite liquor base because of an ample and inexpensive supply of limestone (calcium carbonate). Today, the use of calcium as a sulfite base has become essentially nonexistent in the United States because the spent liquor from this base is difficult and expensive to recover or burn. As a result, sulfite production has declined relative to the growth in kraft production. At most calcium-based sulfite mills, the process has been altered to use a soluble chemical base (magnesium, ammonia, or sodium) that permits recovery of spent liquor.

Sulfite pulps are associated with the production of many types of paper, including tissue and writing papers. In combination with other pulps, sulfite pulps have many applications. In addition, dissolving pulps (i.e., the highly purified chemical cellulose used in the manufacture of rayon, cellophane, and explosives) were produced solely by use of the sulfite process for many years.

Semichemical. The neutral sulfite semichemical pulping process requires cooking wood chips in a neutral solution of sodium sulfite and sodium carbonate or sodium hydroxide. The combination of the cooking time and concentration of chemical solution is such that only a portion of the lignin is removed, after which the pulp is further modified by mechanical disintegration. Pulp yields of 60 percent to 80 percent are obtained, compared with yields of 45 percent to 50 percent for the full chemical process.

This process is acceptable because of its ability to use the vast quantities of inexpensive hardwoods previously considered unsuitable for producing low-quality pulp. Also, the quality of stiffness that hardwood semichemical pulps impart to corrugated board and the large demand for corrugating have promoted a rapid expansion of the process. Often, the process is used in mills that also use the kraft process.

Secondary Fibers. The degree of processing secondary fibers depends on their quality and the product being produced. Some wastepaper (paper stock in industry jargon) can be used with little or no preparation, particularly if the wastepaper is purchased directly from converting operations where a similar product grade is manufactured. At mills where low-quality paper products (e.g., industrial tissue, coarse consumer tissue, molded items, builders' papers, and many types of paperboard) are made, extensive use is made of wastepaper as the raw material. Manufacture of higher-quality products, such as sanitary tissue and printing papers, may involve the use of small percentages of wastepaper. These products require clean, segregated wastepaper and a more extensive preparation system, usually including a de-inking system.

Bleaching of Wood Pulps

After pulping, the unbleached pulp is brown or deeply colored because of the presence of lignins and resins or because of inefficient washing of the spent cooking liquor from the pulp. In order to remove these color bodies from the pulp and to produce a light-colored or white product, it is necessary to bleach the pulp. The brightness of bleached pulp is a function of the raw material and pulping process. Pulps from softwood kraft processes are about brightness 25. Some unbleached sulfite pulps are in the 50-55 range.

The degree of pulp bleaching for paper manufacture is measured in terms of units of brightness and is determined optically using methods established by the Technical Association of the Pulp and Paper Industry (TAPPI). Partly bleached pulps (semi-bleached) are used in making newsprint, food containers, computer cards, and similar papers. Fully bleached pulp is used for white paper products. By bleaching to different degrees, pulp of the desired brightness can be manufactured up to a level of at least 92 on the brightness scale of 100.

Bleaching is normally performed in several stages in which different chemicals are applied. The following symbols are commonly used to describe a bleaching sequence:

Symbol	Bleaching Chemical or Step Represented by Symbol
A	Acid Treatment or Dechlorination
C	Chlorination
D	Chlorine Dioxide Addition
E	Alkaline Extraction
H	Hypochlorite Addition
HS	Hydrosulfite Addition
O	Oxygen Addition
P	Peroxide Addition
PA	Peracetic Acid Addition
W	Water Soak
( )	Simultaneous Addition of the Respective Agents
/	Successive Addition of the Respective Agents Without Washing in Between

The symbols can be used to interpret bleaching "shorthand," which is used in later sections of this report. For example, a common sequence in kraft bleaching, CEDED, is interpreted as follows:

C = chlorination and washing,
E = alkaline extraction and washing,
D = chlorine dioxide addition and washing,
E = alkaline extraction and washing,
D = chlorine dioxide addition and washing.

Almost all sulfite pulps are bleached, but usually a shorter sequence, such as CEH, is sufficient to obtain the brightness needed because sulfite pulps generally contain lower residual lignin. This sequence involves chlorination, alkaline extraction, and hypochlorite application, each followed by washing.

Mechanical pulps (i.e., groundwood) contain essentially all of the wood substrate, including lignin, volatile oils, resin acids, tannins, and other chromophoric compounds. The use of conventional bleaching agents would require massive chemical dosages to enable brightening to levels commonly attained in the production of bleached fully cooked kraft or sulfite pulps. Generally, mechanical pulps are less resistant to aging because of the resin acids still present, and are used in lower-quality, short-life paper products, such as newsprint, telephone directories, catalogs, or disposable products. For these products, a lower brightness is acceptable. Groundwood may be used as produced, at a brightness of about 58 to the mid-60s (GE brightness), or may be brightened slightly by the use of sodium hydrosulfite, sodium peroxide, or hydrogen peroxide. Generally, a single application in one stage is used, but two stages may be used if a higher brightness is required.

Secondary fibers are often bleached to meet the requirements of specific paper products. Again, the choice of bleaching sequence depends on whether the processed stock is composed of only fully bleached chemical pulps or appreciable amounts of groundwood. If the latter, more bleaching will be required, depending on the product being produced; however, bleach demand is usually minimal compared to that in a pulp mill bleachery.

Papermaking

The papermaking process is generally the same, regardless of the type of pulp used or the end product produced. A layer of fiber is deposited from a dilute water suspension of pulp on a fine screen, called the "wire," which permits the water to drain through and retains the fiber layer. This layer is then removed from the wire, pressed, and dried.

Two basic types of paper machines and variations thereof are commonly employed. One is the cylinder machine in which the wire is on cylinders that rotate in the dilute furnish, and the other is the fourdrinier in which the dilute furnish is deposited upon an endless wire belt. Generally, the fourdrinier is associated with the manufacture of newsprint, consumer products, linerboard, and strength papers, and the cylinder machine is used with paperboard production. The fourdrinier has evolved in recent years with such innovations as twin wire formers and vertiformers. The basic objective in the evolution is to produce a more uniform sheet more efficiently. The highest speed of paper machines has now reached at least 5,000 feet per minute.

Converting

The output from the papermaking operation is a "jumbo roll" or stacks of cut sheets. This bulk output is then cut, formed, and/or packaged to yield the final paper product for use -- e.g., package of paper towels, corrugated containers, grocery bags, reams of typing paper. Converting is a fairly energy-intensive process. Except for a few "final" products, such as consumer products (e.g., tissues, napkins, towels, and sometimes bags and typing paper), converting operations to produce paper products for use are not done at the paper mill, but rather at separate "converting" plants.

SOURCES OF WASTES

The quantities and types of wastes generated in the manufacture of paper products are a function of the type of raw material, including species of wood, the pulping process, the degree of bleaching and bleaching sequence, the papermaking process, the converting operation, and the characteristics desired in the final product -- e.g., brightness, wet strength, opacity. To produce a final product with given characteristics, various combinations of the other variables are possible. The degree of whiteness desired is a particularly important variable because the bleaching operation is both energy intensive and the generator of major amounts of liquid wastes. Figure A-2 also illustrates the sources of wastes in the manufacture of paper products, with particular reference to liquid wastes.

Water Effluents

With the exception of dry debarking, water is used in all major operations in pulp and paper mills -- wood preparation, pulping, bleaching, and papermaking. It is used as a medium of transport, a cleaning agent, and a solvent or mixer.

Three categories of water pollutants result from the operations of pulp and paper mills -- conventional, toxic, and nonconventional. The five conventional pollutant parameters are biochemical oxygen demand (BOD), total suspended solids (TSS), pH, fecal coliform, and oil and grease. Toxic pollutants are any of the 129 toxic pollutants listed by EPA [1982]. Nonconventional pollutants are any of 14 additional pollutants. Based on a literature review and extensive sampling at approximately 100 mills, EPA determined that only some of the conventional and toxic pollutants parameters were of current regulatory significance in wastewater discharge from pulp and paper mills. Those pollutant parameters are:

Conventional Pollutants	BOD
	TSS
	pH

<u>Toxic</u>   tricholorophenol
<u>Pollutants</u>   pentachlorophenol
                zinc

The only liquid wastes modeled in this report are the conventional pollutants -- BOD and TSS. BOD, in the form of dissolved organic and inorganic solids, is generated primarily in the pulping and bleaching operations. BOD includes various chlorinated hydrocarbons formed in bleaching with chlorine. TSS is generated primarily in wood preparation, pulp washing, and paper making (white water). TSS includes portions of the wet additives that are applied.

The two conventional water pollutants modeled in this study, BOD and TSS, vary by production subcategory and degree of pollution control. Table A-2 shows changes in unit effluent factors per unit of production for three illustrative subcategories. To reduce discharge, pulp and paper mills use both internal measures related to specific processes and external treatment processes related to aggregate mill effluent. Internal controls represent effluent discharge after the installation of economically justified process controls, but with no end-of-pipe wastewater treatment. Primary treatment effluent factors characterize industry practices before implementation of the Clean Water Act of 1972. These factors reflect the installation of primary sedimentation that reduced gross discharge of solids (approximately 60 to 80 percent) and some BOD (approximately 30 to 35 percent). They represent effluents that occurred in the absence of federal regulation. The biological treatment effluent factors characterize industry practices in the mid-1980s. These factors reflect the Best Practicable Technology effluent guidelines established by EPA between 1976 and 1984 for existing mills. They were more stringent on average than the effluent limits in permits issued by states. The upgraded biological treatment controls represent more stringent effluent limits that industry could feasibly meet by installing additional wastewater treatment equipment. The factors reflect EPA deliberations about Best Conventional Technology guidance [EPA. 1986].

As can be seen in Table A-2, there is considerable variation in waste discharge by subcategory, after both installation of internal controls and biological treatment, primarily reflecting the differences in raw material modification. For example, the bleached kraft process removes more lignin from the pulp than the groundwood process. The lignin reduction results in a BOD effluent factor from internal controls that is double that for groundwood. The sulfite process results in a larger BOD effluent factor

Table A-2. Unit Pollutant Factors by Illustrative Subcategory and
Pollution Control (kg per kkg of production)

	Water Pollution							
	Internal Controls		Primary Treatment		Biological Treatment		Upgraded Biological Treatment	
Subcategory	BOD	TSS	BOD	TSS	BOD	TSS	BOD	TSS
Bleached Kraft	39	71	39	24	4	8	3	5
Sulfite	135	91	135	16	11	17	5	6
Groundwood	19	71	19	24	2	8	1	4

	Air Pollution								
	Baseline			Existing			New		
Subcategory	TSP	SO2	TRS	TSP	SO2	TRS	TSP	SO2	TRS
Bleached Kraft									
process	8	21	14	3	21	1	2	21	neg
boiler (wood)	5	neg	-	1	neg	-	neg	-	
boiler (coal)	24	14	-	2	12	-	1	6	-
Sulfite									
process	1	118	-	1	13	1	neg	-	
boiler (wood)	6	neg	-	neg	-	neg	-		
boiler (coal)	27	15	-	neg	14	-	neg	7	-
Groundwood									
process	-	-	-	-	-	-	-	-	-
boiler (wood)	-	-	-	-	-	-	-	-	-
boiler (coal)	23	13	-	2	12	-	neg	6	-

	Solid Waste[a]			
Subcategory	Internal Controls	Primary Treatment	Biological Treatment	Upgraded Biological Treatment
Bleached Kraft	6	53	81	85
Sulfite	0	74	114	127
Groundwood	0	40	52	54

a. Sludge generation. Figures are cumulative.
Source: E.C. Jordan [1984].

than the kraft process primarily because fewer sulfite mills have recovery systems with the same level of recovery as is typical for kraft mills.

Air Emissions

Air pollutants in pulp and paper production come from various production processes and from power boilers used to generate steam and/or electricity. The pulping process and the chemical recovery operation in kraft mills are major sources of process emissions. All mills use power boilers to generate process steam, and many mills have power boilers to generate electric energy. Fuel combustion in power boilers is a major source of emissions.

Three categories of air pollutants result from the operation of pulp and paper mills -- criteria, hazardous, and designated. The criteria air pollutants are the six gases and solids identified in the Clean Air Act. Hazardous air pollutants are toxic emissions that have adverse health effects. Designated air pollutants are emissions that cause either adverse health or welfare effects but are not classified as either criteria or hazardous. Air pollutants of current regulatory significance from pulp and paper mill production processes and power boilers are as follows:

Category of Air Pollutant	Air Pollutants	
	Production Processes	Power Boilers
Criteria	TSP and SO_2	TSP, SO_2, NO_x, CO
Designated	TRS	none
Hazardous	none	none

The air pollutants modeled in this study vary by production process, boiler fuels, and degree of air pollution control. Table A-2 shows changes in unit emission factors per unit of production for three illustrative subcategories. The baseline emission factors characterize industry practices before the implementation of the Clean Air Act of 1970. These emission factors reflect uncontrolled emissions or, in some cases, such as TSP, emissions after installation of pollution control equipment for economic or process reasons. They represent emissions that occurred in the absence of any state or federal regulation. The existing emission factors characterize industry practices in the mid-1980s. These factors reflect current emission regulations specified in State Implementation Plans for existing plants. The new source controls characterize more stringent emission levels that industry could feasibly meet by installing additional controls on processes and boilers. These levels reflect EPA's New Source Performance Standards regulations for pulp and paper mills and power boilers.

Bleached kraft pulping has several process sources with air emissions, including the digester/evaporator, brown stock washer, recovery furnace, smelt dissolving tank, and lime kiln. Particulates are emitted largely from the recovery furnace, the lime kiln, and the smelt dissolving tank. The characteristic odor of kraft pulping is caused by the emission of reduced sulfur compounds, the most common of which is hydrogen sulfide. The major source of hydrogen sulfide is the direct-contact evaporator, used with older recovery boilers in which the sodium sulfide in the black liquor reacts with the carbon dioxide in the furnace exhaust. The other compounds are emitted from several points within a mill, but the main sources are the digester/blow tank systems and the direct-contact evaporator.

Sulfite pulping has several unique processes, such as acid tower and screens. Sulfur dioxide emissions are generally considered the major pollutant of concern. The major source is the digester and blow pit (dump tank) system. Another source is the recovery system. The only significant source of particulate emissions is the recovery furnace exhaust. The characteristic "kraft" odor is not emitted because volatile reduced sulfur compounds are not products of the lignin/bisulfite reaction.

Although boiler emissions associated with different pulping processes do not vary significantly, they do exhibit considerable variation by fuel use. For example, unit TSP emissions from coal-fired boilers at bleached kraft mills are five times greater than from wood-fired boilers. Whereas residual oil and natural gas were the primary sources of fuel in the 1970s, wood and coal are the primary sources in the 1980s. Consequently, the potential for TSP and SO_2 emissions from the industry increased significantly between the early 1970s and the mid-1980s.

Solid Wastes

Solid wastes come from production processes, wastewater treatment facilities, and air pollution control systems. Organic pollutants characterize solid wastes generated by water pollution controls, whereas metals characterize solid wastes generated by air production processes. Both sludge and dry solids are typically disposed of in unlined landfills.

The solid wastes from pulp and paper mills are classified as conventional and toxic wastes. Conventional wastes are dry solids and primary and secondary wastewater treatment sludges that are free of significant levels of toxic chemicals. Toxic wastes are those with significant levels of toxic chemicals.

The solid wastes generated by the industry (and not modeled in this study) are separated into those generated by production processes (process solids), by water pollution controls (sludge), by air pollution controls (e.g., collected fly ash), and by power boilers (boiler ash). We assume that all process wastes recovered by air pollution control systems are recycled and therefore do not add to solid wastes disposed of in landfills. However, for many types of waste, this is not possible or practical (e.g., fly ash).

The cumulative amount of solid waste per unit of production from wastewater controls is delineated in Table A-2. The baseline factors characterize industry practices before implementation of the Clean Water Act of 1972. The existing control factors characterize industry practices in the mid 1980s as a result of meeting Best Practicable Technology effluent guidelines. The new source control factors would characterize industry practices if EPA were to require installation of more stringent conventional pollution control.

ESTIMATED GENERATION OF WASTES

Table A-3 summarizes the estimates of discharges of liquid, gases, and solids from the pulp and paper industry for 1973 and 1984. As one would have anticipated, liquid and gaseous effluents declined over the period. Solid effluents increased because of the removal of solids from liquid effluents.

Table A-3. Estimated Discharges to the Environment by the
Pulp and Paper Industry[a] -- 1973 and 1984 (10^6 tons)

	1973	1984
Liquid Effluents		
BOD_5	0.70	0.20
TSS	0.68	0.19
Gaseous Effluents		
TSP	0.35	0.16
SO_2	0.77	1.75
TRS	0.06	0.01
Solid Effluents		
Sludge, dry solids basis	1.95	4.0
Process solids	4.5	5.8
Collected fly ash & boiler ash	NA	NA

a. Excludes converting only operations.
Source: Chapter 3 for liquid effluents, Chapters 4 and 5 for gaseous effluents,
and E.C. Jordan [1984] for solid effluents.

Note

1. We projected the estimated 1994 capacity in 1986. More current
information suggests that pulp capacity will increase by 12,000 rather than
8,100 tons and that paper and paperboard capacity will increase by 17,000
rather than 9,000 tons [API. 1989].

References

American Paper Institute (API). 1974. *Paper Paperboard and Wood Pulp
Capacity.* New York, NY.

_____. 1985. *Paper Paperboard and Wood Pulp Capacity.* New York,
NY.

_____. 1987. *Paper Paperboard and Wood Pulp Capacity.* New York,
NY.

_____. 1989. *Paper Paperboard and Wood Pulp Capacity.* New York,
NY.

Brandes, Debra A. (ed.) 1984. *1985 Post's Pulp and Paper Directory.* Miller Freeman Publications, San Francisco, CA.

Jordan, E.C. 1984. "Pollution Analysis." In Putnam Hayes & Bartlett, Inc., "Analysis of Cost Effective Pollution Control Strategies in the Pulp & Paper Industry," Volume II, Appendix C. Report to the Office of Policy, Planning, and Evaluation, U.S. Environmental Protection Agency. Cambridge, MA.

Pulp and Paper International (PPI). 1985. *International Pulp and Paper Directory.* Miller Freeman, San Franciso, CA.

U.S. Environmental Protection Agency (EPA). 1982. "Development Document for Effluent Limitations and Guidelines and Standards for the Pulp, Paper, and Paperboard and Builders' Paper and Board Mills." EPA-440/1-82/025. Washington, DC.

_____. 1986. "Development Document for Best Conventional Pollutant Control Technology Effluent Limitation Guidelines and Standards for the Pulp, Paper and Paperboard and Builders' Paper and Board Mills Point Source Categories." EPA-440/1-86/025. Washington, DC.

Appendix B

MILLS EVALUATED IN THE STUDY

Table B-1. Mills Evaluated in the Study

Mill	ST	Mill Type D/I	S/K/G	Mills Included in 1984 Analysis Water	TSP	TRS	TOX	CHCL$_3$	Haz
Alaska Lumber & Pulp	AK	D	S		X	X	X		
Louisiana Pacific Corp	AK	D	S				X		
Alabama River Pulp Co	AL	D	K	X		X		X	
Allied Paper	AL	D	K	X	X	X	X	X	
Champion International	AL	D	K	X	X	X	X	X	X
Container Corp of America	AL	D	K	X	X	X	X		
Georgia Kraft	AL	D	K	X	X	X			
Gulf States Paper	AL	D	K	X	X	X	X	X	X
Hammermill Papers Group	AL	D	K	X	X	X	X		X
International Paper	AL	D	K	X	X	X	X		
James River Corp.	AL	D	K	X	X	X	X	X	X
Kimberly Clark Corp.	AL	D	K	X	X	X	X		
MacMillan Bloedel	AL	D	K	X	X	X			
Mead	AL	D	S	X	X				
National Gypsum	AL	D	G	X	X				
Scott Paper	AL	D	K	X	X	X	X	X	
Stone Container	AL	D		X					
Union Camp Corp.	AL	D	K	X	X	X			
Arkansas Kraft	AR	D	K	X	X	X			
Bear Brand Roofing Inc.	AR	I		X					
Celotex Corp.	AR	I		X					
Elk Corp.	AR	D		X					
Genstar Blgd. Mat.Co.	AR	I		X					
Georgia Pacific	AR	D	K	X	X	X	X	X	
International Paper (Pine Bluff)	AR	D	K	X	X	X	X	X	
International Paper (Camden)	AR	D	K	X	X	X			
Nekoosa Papers	AR	D	K	X	X	X	X	X	
Potlatch	AR	D	K	X	X	X	X	X	
Weyerhaueser	AR	D	K	X	X	X			
Southwest Forest Ind.	AZ	D	K	X	X		X		
B J Fibres Inc.	CA	I		X					
California Paperboard Co.	CA	I		X					
Cellulo Co.	CA	I		X					
Celotex Corp.	CA	I		X					
Certainteed Corp.	CA	I		X					
Container Corp. (Los Angeles)	CA	I		X					
Container Corp. (Santa Clara)	CA	I		X					
Crown Zellerbach Corp	CA	D	K	X					
Diamond International	CA	D		X					
Domtar Gypsum Amer Inc.	CA	I		X					
Ped Paper Board Co. Inc.	CA	I		X					
Garden St. Paper Co Inc.	CA	I		X					

(cont.)

Table B-1 (cont.). Mills Evaluated in the Study

Mill	ST	Mill Type D/I	S/K/G	Mills Included in 1984 Analysis Water	TSP	TRS	TOX	CHCL$_3$	Haz
Genstar Bldg. Mat. Co.	CA	I		X					
Inland Container Corp.	CA	I		X					
Keyes Fibre Co.	CA	I		X					
Kimberly Clark Corp.	CA	I		X					
LA-Pacific/Fibreboard	CA	D	K	X	X	X		X	
Leatherback Industries	CA	I		X					
LA Paper Box & Boar	CA	I		X					
Louisiana-Pacific	CA	D	K		X	X		X	
Owens-Corning Fiberglass	CA	I		X					
Pac. Coast Pack. Corp	CA	I		X					
Paper Pak Products Inc.	CA	I		X					
Potlatch Corp.	CA	I		X					
Sierra Tissue Inc.	CA	I		X					
Simpson Paper (Fairhaven)	CA	D	K		X	X	X	X	
Simpson Paper (Ripon)	CA	D							
Simpson Paper (Anderson)	CA	D	K	X	X	X	X		X
Sonoco Prod. (City of Industry)	CA	I		X					
Sonoco Prod. (Richmond)	CA	I			X				
US Gypsum Co.	CA	I		X					
Willamette Industries	CA	I		X					
Packaging Corp of Am.	CO	I		X					
Kimberly Clark	CT	D		X					
Ludall & Foulds	CT	I		X					
Robertson Paper Bos Co	CT	I		X					
Rodgers Corp.	CT	I		X					
Windsor Stevens	CT	D		X					
Container Corp. of Amer.	DE	I		X					
Curtis Paper Co.	DE	I		X					
NVF Co. (Newark)	DE	I		X					
NVF Co. (Yorklyn)	DE	D		X					
Alton Packaging Corp.	FL	D	K	X	X	X			
Buckeye Cellulose	FL	D	K	X	X	X	X		
Container Corp.	FL	D	K	X	X	X			
Georgia Pacific Corp.	FL	D	K	X	X	X	X	X	
ITT Rayonier	FL	D	S	X	X		X		
Jacksonville Kraft	FL	D	K	X	X	X			
Simkins Industries Inc.	FL	I		X					
Southwest Forest Ind.	FL	D	K		X	X	X	X	
St. Joe Paper Co.	FL	I	K	X	X	X	X	X	
St. Regis	FL	D	K	X	X	X	X	X	
US Gypsum Co.	FL	I		X					
Alton Packaging	GA	D		X					
Augusta Newsprint Co.	GA	D	G	X			X		
Austell Boxboard Corp	GA	I		X					
Brunswick Paper & Pulp	GA	D	K	X	X	X	X	X	X
Buckeye Cellulose	GA	D	K				X		
Continental For.(Port Wentworth)	GA	D	K	X	X	X			
Continental For.(Augusta)	GA	D	K	X	X	X		X	X
Deerfield Specialty Prod	GA	D		X			X		
Georgia Kraft (Krannert)	GA	D	K	X	X	X			
Georgia Kraft (Macon)	GA	I	K	X	X	X			
Gilman Paper Co.	GA	D	K	X	X	X	X		
Great Southern Paper	GA	D	K	X	X	X			
Interstate Paper	GA	D	K	X	X	X			

(cont.)

Table B-1 (cont.). Mills Evaluated in the Study

Mill	ST	Mill Type D/I	S/K/G	Water	TSP	TRS	TOX	CHCL$_3$	Haz
ITT Rayonier	GA	D	K	X	X	X	X		
Owens Illinois	GA	D	K	X	X	X			
Owens-Corning Fiberglass	GA	I		X					
Packaging Corp. of Amer.	GA	I		X					
Ponderosa Georgia Corp.	GA	I		X					
Sonoco Products Co.	GA	I		X					
Southeast Paper Mfg.	GA	D		X					
Union Camp	GA	D	K	X	X	X			
Consol. Pack. Corp.	IA	D		X					
Packaging Corp of Am.	IA	D		X					
Potlatch Corp.	ID	D	K	X	X	X	X	X	X
Alton Pack. Corp.	IL	D		X					
Bemis Co. Inc.	IL	I		X					
Container Corp. of Amer.	IL	I		X					
FSC Paper Corp.	IL	I		X					
Georgia Pacific Corp.	IL	D		X					
Johns-Mnaville Corp.	IL	I		X					
Nobisco Brands	IL	I		X					
Prairie State Paper Mill	IL	I		X					
Sonoco Products	IL	D		X					
The Celotex Corp.	IL	D		X					
The Quaker Oats Co.	IL	D		X					
Alton Packaging Corp.	IN	I		X					
Clevepak	IN	D		X					
Cont. Corp. of Amer.	IN	D		X					
Inland Container Corp.	IN	D		X					
Kieffer Paper Mills	IN	D		X					
Lloyd A Frye Roofing	IN	I		X					
Packaging Corp. of Amer.	IN	I		X					
Weston Paper & Manuf.	IN	D		X					
3M Co	IN	I		X					
Packaging Corp. of Amer.	KS	I		X					
Knowlton Brothers Inc.	KY	I		X					
Westvaco	KY	D	K	X	X	X	X		
Willamette Ind. (Hawesville)	KY	D	K	X	X	X		X	
Willamette Ind. (Hawesville)	KY	D	S	X	X	X	X		X
W R Grace	KY	D		X					
Boise Cascade	LA	D	K	X	X	X			
Boise Southern	LA	D	K	X	X	X	X	X	
Continental Forest Ind	LA	D	K	X	X	X			
Crown Zell. (Bogalusa)	LA	D	K	X	X	X			
Crown Zell. (St. Francesville)	LA	D	K	X	X	X	X	X	
Georgia Pacific	LA	D	K	X	X	X	X		X
International Paper (Mansfield)	LA	D	K	X	X	X			
International Paper (Springhill)	LA	D	K	X					
International Paper (Bastrop)	LA	D	K	X	X	X	X	X	
International Paper (Pineville)	LA	D	K	X	X	X			
Manville Forest Products	LA	D	K	X	X	X			
Valentine Pulp and Paper	LA	D	S	X	X				
Willamette	LA	D	K	X	X	X			
Baldwinville Prod. Inc	MA	I		X					
Crane & Co Inc	MA	D		X					
Crocker Tech Papers Inc	MA	I		X					
Deerfield Spec Pap Inc	MA	D		X					

(cont.)

Table B-1 (cont.). Mills Evaluated in the Study

Mill	ST	Mill Type		Mills Included in 1984 Analysis					
		D/I	S/K/G	Water	TSP	TRS	TOX	CHCL$_3$	Haz
Erving Paper Mills	MA	D		X					
Esleek Mfg	MA	D		X					
Paper Co	MA	I		X					
Millers Falls	MA	I		X					
Turners Falls	MA	D		X					
Haverhill Papbrd Corp	MA	I		X					
Hollings. & Vose (West Groton)	MA	D		X					
Hollings. & Vose (East Walpole)	MA	I		X					
James River (Adams)	MA	I		X					
James River (Fitchburg)	MA	I		X					
James River (E. Pepperell)	MA	D		X					
James River/Mass (Fitchburg)	MA	D		X					
Kimberly Clark	MA	D		X					
Lawrence Paperboard Co	MA	I		X					
Linweave Inc	MA	I		X					
Mead Corp/Laurel Mill	MA	D		X					
Mead Corp/Willow	MA	D		X					
Merrimac Paper Co Inc	MA	I		X					
Natick Paperboard Corp	MA	I		X					
NVF/Parson's Paper	MA	D		X					
Parsons Paper	MA	I		X					
Perkit Folding Box Corp	MA	I		X					
Premoid Corp	MA	I		X					
Rising Paper Co	MA	I		X					
Seaman Paper Co/Mass	MA	D		X					
Southworth Co	MA	I		X					
Sunoco Prod.	MA	I		X					
Texon Inc (South Hadley)	MA	I		X					
Texon Inc (Russell)	MA	D		X					
Texon Inc (South Hadley)	MA	D		X					
Westfield River Paper	MA	D		X					
Chesapeake Paperboard Co	MD	I		X					
Simkins Industries Inc	MD	I		X					
Westvaco	MD	I	K		X	X		X	
Boise Cascade	ME	D		X	X	X		X	X
Eastern Fine Paper Inc	ME	I		X					
Fraser Paper Inc	ME	D		X					
Georgia Pacific Corp	ME	D	K	X	X	X	X	X	
Great Northern (E. Millinocket)	ME	D	G	X	X				
Great Northern (Millinocket)	ME	D	S	X	X				
International Paper Co	ME	D	K	X	X	X	X		X
James River Corp	ME	D	K	X	X		X	X	X
Keyes Fibre Co	ME	I	G	X					
Lincoln Pulp/Paper Inc	ME	D	K	X	X	X	X		
Madison Paper Industries	ME	I	G	X					
Pejepscot Paper	ME	D		X					
S.D. Warren (Westbrook)	ME	D	K	X	X	X	X	X	X
Scott Paper	ME	D		X					
SD Warren (Hinckley)	ME	D		X			X		
St Regis Corp	ME	D	G	X					
Statler Tissue	ME	D		X			X		
Yorktowne Paper (Gardner)	ME	D		X					
Yorktowne Paper (Gardner)	ME	I		X					
Allied Paper Inc	MI	D		X					

(cont.)

Table B-1 (cont.). Mills Evaluated in the Study

Mill	ST	Mill Type		Mills Included in 1984 Analysis					
		D/I	S/K/G	Water	TSP	TRS	TOX	CHCL₃	Haz
Big M Paperboard	MI	D		X					
Champion Intrntl Corp	MI	D		X					
Dunn Paper	MI	D		X					
Fletcher Paper Co	MI	I		X					
French Paper	MI	D		X					
Georgia Pacific Corp	MI	I		X					
James River (Kalamazoo)	MI	D		X					
James River (Kalamazoo)	MI	I		X					
James River Rochester	MI	I		X					
Kimberly Clark Corp	MI	D		X					
Manistique Papers Inc.	MI	D	G	X	X				
Mead Corp (Otsego)	MI	D		X	X	X			
Mead Corp (Escanaba)	MI	D	K	X	X	X	X		X
Menasha Corp	MI	D		X					
Menominee Paper Co Inc	MI	I		X					
National Gypsum Co	MI	I		X					
Packaging Corp of Amer	MI	D		X					
Peninsular Paper Co	MI	I		X					
Plainwell Paper	MI	D		X					
Port Huron Paper	MI	D		X					
Proctor & Gamble Paper	MI	D		X					
S.D Warren Co	MI	I	K	X	X	X	X	X	
Scott Paper Co	MI	I		X					
Simplex Industries Inc	MI	I		X					
Simplicity Pat Co Inc	MI	D		X					
Smurfit Paperboard	MI	I		X					
Union Camp Corp	MI	I		X					
Blandin Paper Co	MN	I	G	X	X				
Boise Cascade	MN	D	K	X	X	X	X	X	X
Certainteed Corp	MN	I		X					
Champion International	MN	I		X					
Hennepin Paper	MN	D	G	X			X		
Potlatch (Brainerd)	MN	I		X					
Potlatch (Cloquet)	MN	D	K	X	X	X	X	X	
St Regis Corp	MN	D	G	X					
Superwood Corp	MN	I	G						
Tamko Asphalt Prod Inc	MO	I		X					
Atlas Roofing	MS	I	G		X	X			
Burrows Southern Corp	MS	D		X					
Diamond Intrntl Corp	MS	D		X					
Dunn Paper Co	MS	D		X					
International Paper (Natchez)	MS	D	K	X	X	X	X		
International Paper (Moss Point)	MS	D	K	X	X	X	X	X	
International Paper (Vicksburg)	MS	D	K	X	X	X			
St Regis	MS	D	K	X	X	X			
Weyerhaeuser	MS	D		X					
Champion Int. Packaging	MT	D	K		X	X	X	X	
Alpha Cellulose Corp	NC	D		X					
Cellu Products	NC	D		X					
Champion Int. (Roanoke Rapids)	NC	D	K	X	X	X			
Champion Int. (Canton)	NC	D	K	X	X	X	X	X	
Federal Paperboard Co	NC	D	K	X	X	X	X	X	X
Halifax Paperboard Inc	NC	I		X					
Olin Corp	NC	D		X					

(cont.)

Table B-1 (cont.). Mills Evaluated in the Study

Mill	ST	Mill Type		Mills Included in 1984 Analysis					
		D/I	S/K/G	Water	TSP	TRS	TOX	CHCL$_3$	Haz
Weyerhaeuser (New Bern)	NC	D	K	X	X	X	X	X	X
Weyerhaeuser (Plymouth)	NC	D	K	X	X	X	X	X	
Ashuelot Paper	NH	D		X					
Brown Products Inc	NH	I		X					
Coy Paper	NH	D		X					
CPM Corp	NH	D		X					
Groveton Papers	NH	D	S	X	X				
Hinsdale Products Co	NH	D		X					X
Hoague Sprague Corp	NH	D		X					
James River	NH	D	K	X	X	X	X		
James River/Brown	NH	D		X					
Lydall Inc	NH	I		X					
Monadnock Paper Mills	NH	D		X					
Paper Service Mills	NH	D		X					
Spaulding Fibre Co Inc	NH	D		X					
Celotex Corp (Linden)	NJ	I		X					
Celotex Corp (Perth Amboy)	NJ	I		X					
Davey Co	NJ	I		X					
Garden St Paper Co Inc	NJ	I		X					
Georgia Pacific Corp	NJ	D		X					
James River (Hughesville)	NJ	D		X					
James River (Milford)	NJ	D		X					
James River (Warren Glen)	NJ	D		X					
Johns-Manville Corp	NJ	D		X					
Kimberly Clark Corp	NJ	I		X					
Lowe Paper Co	NJ	I		X					
Marcal Paper Mills Inc	NJ	I		X					
Newark Group Inc	NJ	I		X					
US Gypsum (Camden)	NJ	I		X					
US Gypsum (Clark)	NJ	I		X					
Armstrong World Ind Inc	NY	D		X					
Bio Tech	NY	D		X					
Boise Cascade/Latex	NY	D		X					
Boise Cascade/Lewis	NY	D		X					
Brownville Paper	NY	I		X					
Burrows Paper (Lyonsdale)	NY	D		X					
Burrows Paper (Little Falls)	NY	I		X					
Chagrin Fibres Inc	NY	D		X					
Climax Mfg	NY	I		X					
Columbia Corp (Chatham)	NY	D		X					
Columbia Corp (N. Hoosick)	NY	D		X					
Cornwall Paper Mill	NY	D		X					
Cottrell Paper Co Inc	NY	D		X					
Crown Zell. (S. Glen Falls)	NY	D		X					X
Crown Zell. (Carthage)	NY	D		X					
Crown Zell. (Carthage)	NY	I		X					
Finch, Pruyn & Co	NY	D	S	X	X		X	X	
Flower Cty Tissue Mls	NY	I		X					
Fort Orange Paper Co	NY	I		X					
Foster Paper Co	NY	I		X					
Georgia Pacific (Plattsburgh)	NY	I		X					
Georgia Pacific (Lyons Falls)	NY	D		X					
Hammermill Paper Co	NY	I		X					
Hollingsworth & Vose	NY	D		X					

(cont.)

Table B-1 (cont.). Mills Evaluated in the Study

Mill	ST	D/I	S/K/G	Water	TSP	TRS	TOX	CHCL$_3$	Haz
Hollingsworth & Vose	NY	D		X					
Imperial Paper Co	NY	I		X					
International Paper (Corinth)	NY	D	K	X	X	X			
International Paper (Ticonderoga)	NY	D		X			X		X
James River	NY	D		X					
Kimberly Clark Corp	NY	D		X					
Koppers Co	NY	I		X					
Lydall Inc	NY	D		X					
Manning Paper	NY	I		X					
McIntyre Paper Co Inc	NY	D		X					
Mohawk Paper Mills	NY	I		X					
Mohawk Paper Mills	NY	D		X					
Mohawk Valley Paper Co	NY	I		X					
Newton Falls Paper Mills	NY	D		X					
Norton Pulp & Mach Inc	NY	D							
Potsdam Paper Corp	NY	D		X					
Red Hook Paper Corp	NY	I		X					
Rock-Tenn	NY	D		X					
Schoeller Technl Papers	NY	D		X					
Scott Paper	NY	D		X					
Sonoca Prod.	NY	I		X					
Spaulding Fibre Co	NY	I		X					
St Regis Corp	NY	D	G	X	X				
Stevens & Thompson Paper	NY	D		X					
Tagsons Paper	NY	I		X					
Upson Co	NY	I		X					
US Gypsum	NY	D		X					
Beckett Paper Co	OH	I		X					
Celotex Corp	OH	I		X					
Certainteed	OH	D		X					
Champion Internl Corp	OH	I		X					
Chase Bag Co	OH	D		X					
Cheney Pulp & Paper Co	OH	I		X					
Container Corp. (Cincinnati)	OH	I		X					
Container Corp. (Circleville)	OH	D		X					
Crown Zellerbach	OH	D		X					
Crystal Tissue	OH	D		X					
Erving Paper Mills	OH	I		X					
Harding-Jones Paper	OH	D		X					
Howard Paper Mills Inc	OH	I		X					
Mead Corp (Cincinnati)	OH	I		X					
Mead Corp (Chillicothe)	OH	D	K	X	X	X	X	X	X
Miami Paper	OH	D		X					
Middletown Paperboard Co	OH	I		X					
Mosinee Paper Corp	OH	I		X					
Packaging Corp of Ameri	OH	D		X					
PH Glatfelter	OH	D		X					
Smurfit Diamond Pkg	OH	I		X					
Sorg Paper Co	OH	D							
St Regis (Columbus)	OH	D		X					
St Regis (Columbus)	OH	I		X					
Stone Container (Franklin)	OH	D	K	X		X			
Stone Container (Franklin)	OH	I		X					
Sunoco Prod.	OH	I		X					

(cont.)

Table B-1 (cont.). Mills Evaluated in the Study

Mill	ST	D/I	S/K/G	Water	TSP	TRS	TOX	CHCL$_3$	Haz
Tecumseh Corrugated Box	OH	D		X					
Toronto Paperboard	OH	D		X					
US Gypsum	OH	D		X					
Fort Howard Paper	OK	D		X					X
Georgia Pacific	OK	D							
National Gypsum Co	OK	I		X					
Robel	OK	D							
Weyerhaeuser (Broken Bow)	OK	D							
Weyerhaeuser (Valliant)	OK	D	K	X	X	X			
Bird & Son Inc	OR	I		X					
Boise Cascade Corp	OR	D	K		X	X	X	X	
Concel Inc	OR	I		X					
Crown Zell. (Lebanon)	OR	D	S						
Crown Zell. (West Linn)	OR	D		X					
Crown Zell. (Clatskanie)	OR	D	K	X	X		X	X	
Georgia Pacific	OR	D	K	X	X	X			
International Paper	OR	D	K	X	X	X			
James River Corp	OR	D	K	X	X	X	X	X	X
Malarkey Roofing	OR	I		X					
Publishers Paper (Oregon City)	OR	D	S	X	X				
Publishers Paper (Newberg)	OR	D	S	X	X				
Weyerhaueser	OR	D	K	X	X	X			
Weyerhauser Wt Coast	OR	D	S	X	X				
Willamette	OR	D	K	X	X	X			
American Paper Prod Co	PA	I		X					
Appleton Papers	PA	D	K	X	X	X	X		X
Brandywine Paper Corp	PA	D		X					
Celotex	PA	I		X					
Connelly Containers	PA	I		X					
Container Corp of Ameri	PA	I		X					
Crown Paperboard Co Inc	PA	I		X					
Eaton-Dikeman	PA	D		X					
Georgia Pacific Corp	PA	I		X					
Hammermill Paper (Lock Haven)	PA	D		X					
Hammermill Paper (Erie)	PA	I	K	X	X	X	X	X	
Interstate Intercorr	PA	I		X					
Kimberly Clark	PA	D		X					
National Gypsum	PA	D		X					
Newman & Company Inc	PA	I		X					
Owens-Corning Fiberglas	PA	I		X					
Packaging Corp of Ameri	PA	I		X					
Penntech Papers	PA	D	K	X	X	X	X	X	X
PH Glatfelter	PA	D	K	X	X	X	X	X	X
Potlatch Corp	PA	D		X					X
Proctor & Gamble Paper	PA	D	S	X	X	X	X	X	X
Quin-T Corp	PA	I		X					
Scott Paper Co	PA	I		X					
Simpson	PA	D		X					
Sonoco Prod.	PA	D		X					
St Regis Paper Co	PA	I		X					
Westvaco Corp	PA	I		X					
Woodstream Corp	PA	I		X					
Yorketown Pap Mills Inc	PA	I		X					
Bird & Son Inc	RI	I		X					

(cont.)

Table B-1 (cont.). Mills Evaluated in the Study

Mill	ST	D/I	S/K/G	Water	TSP	TRS	TOX	CHCL$_3$	Haz
Bowater Carolina	SC	D	K	X	X	X	X	X	
International Paper	SC	D	K	X	X	X	X	X	
Kimberly Clark	SC	D		X					
Sonoco Prod.	SC	D	S	X	X				
Stone Container	SC	D	K	X	X	X			
Union Camp	SC	D	K		X	X	X		
Westvaco	SC	D	K	X	X	X			
Bowater Southern Paper	TN	D	K	X	X	X	X	X	
Container Corporation of	TN	I		X					
Harriman Paperboard Corp	TN	I		X					
Inland Container	TN	D	S	X	X				
Knowlton Bros	TN	I		X					
Lydall Inc	TN	D		X					
Mead Corp	TN	D	K	X	X	X	X	X	
National Fibrit	TN	I		X					
Owens-Corning Fiberglas	TN	I		X					
Rock-Tenn Co	TN	I		X					
Sonoco Prod.	TN	D		X					
Tamko Asphalt Prod Inc	TN	I		X					
Tenn River Pulp & Paper	TN	D	K	X	X	X			
Celotex Corp	TX	I		X					
Champion International	TX	I	K	X	X	X	X	X	
Equitable Bag	TX	D		X					
International Paper Co	TX	D	K		X	X	X	X	X
Owens Illinois	TX	D	K	X	X	X			
Rock-Tenn	TX	I		X					
St Regis (Lufkin)	TX	D	K	X	X	X	X	X	
St Regis (Houston)	TX	D	K	X	X	X	X		X
Temple-Eastex	TX	D	K	X	X	X	X	X	
US Gypsum	TX	D		X					
Chesapeake Corp of VA	VA	D	K	X	X	X	X		
Federal Paperboard Co	VA	I		X					
Georgia Bonded Fibers	VA	D		X					
Hercules Inc	VA	I					X		
James River Paper Co	VA	I		X					
Owens Illinois	VA	D	K	X	X	X			
Stone Container Corp	VA	D	K		X	X			
Union Camp	VA	D	K		X	X	X		
Virginia Fibre Corp	VA	D	K	X	X	X			
Westvaco	VA	D	K	X	X	X	X	X	
Bemis Co Inc	VT	D		X					
Boise Cascade	VT	D		X					
Ehv-Weidmann Industries	VT	D		X					
Georgia Pacific	VT	D		X					
Mountain Paper Prod Corp	VT	I		X					
Putney Paper	VT	D		X					X
Standard Packaging	VT	D		X					
Boise Cascade (Vancouver)	WA	D		X					
Boise Cascade (Steilacoom)	WA	D		X					
Boise Cascade (Wallula)	WA	D	K	X	X	X	X		
Container Corp of Amer.	WA	I		X					
Crown Zell. (Camas)	WA	D	S						
Crown Zell. (Port Angeles)	WA	D	G		X	X			
Crown Zell. (Port Angeles)	WA	D	G	X	X	X			

(cont.)

Table B-1 (cont.). Mills Evaluated in the Study

Mill	ST	Mill Type		Mills Included in 1984 Analysis					
		D/I	S/K/G	Water	TSP	TRS	TOX	CHCL$_3$	Haz
Crown Zell. (Camas)	WA	D	K	X	X		X	X	
Georgia Pacific Corp	WA	D	S	X	X		X	X	X
Grays Harbor Paper Co	WA	I		X					
Handle Paper	WA	D	K	X	X	X			
Inland Empire Paper Co	WA	D		X					
Itt Rayonier (Port Angeles)	WA	D	S	X	X		X		
Itt Rayonier (Hoquiam)	WA	D	S	X	X		X		
Keyes Fibre	WA	D		X					
Longview Fibre	WA	D	K	X	X	X	X	X	
North Pacific Paper	WA	D							
Scott Paper	WA	D	S	X	X		X	X	X
Sonoco Prod.	WA	D		X					
St Regis	WA	D	K	X	X	X	X	X	
Weyerhaeuser (Cosmopolis)	WA	D	S	X	X		X		
Weyerhaeuser (Everett)	WA	D	K		X	X	X	X	
Weyerhaeuser (Longview)	WA	D	K	X	X	X	X	X	
Amricon Corp	WI	I		X					
Appleton Papers	WI	D		X					
Badger Paper mills inc	WI	I	S	X	X		X	X	
Beloit Box Board Co	WI	I		X					
Bermico Co	WI	D		X					
Consolidated (Whiting)	WI	D		X					
Consolidated (Appleton)	WI	D	G	X	X				
Consolidated (Stevens Point)	WI	D		X					
Consolidated (Wisconsin Rapids)	WI	D	K	X	X	X	X		
Consolidated (Wisconsin Rapids)	WI	D	G		X				
Filter Materials Inc	WI	I		X					
Flambeau Paper	WI	D	S	X	X		X	X	X
Fort Howard	WI	D		X					X
Fox River Paper Co	WI	I		X					
Genstar Build Materials	WI	D		X					
Georgia Pacific	WI	D		X					
Gilbert Paper Co	WI	I		X					
Green Bay Packaging	WI	D		X					
Hammermill Papers	WI	D	K	X	X	X			
James River	WI	D	S	X	X		X	X	X
James River-Dixie North	WI	I		X					
Kimberly-Clark/Badger Gl	WI	D		X					
Kimberly-Clark/Lakeview	WI	D		X					
Menasha Corp	WI	I		X					
Midtech Paper Corp	WI	D	G	X	X				
Mosinee Paper	WI	D	K	X	X	X			
Neenah Paper Co	WI	D		X					
Neenah Paper/Whiting Mill	WI	D	S	X	X		X	X	
Nekoosa Paper (Port Edwards)	WI	D		X					
Nekoosa Paper (Nekoosa)	WI	D	K	X	X	X	X		
Naigara of Wisc Pap Corp	WI	D	G		X				
Owens, Illinois	WI	D	K	X	X	X			
PH Glatfelter	WI	D		X					
Ponderosa Pulp Prod Corp	WI	I		X					
Pope & Talbot (Eau Claire)	WI	D		X					X
Pope & Talbot (Ladysmith)	WI	D		X					X
Proctor & Gamble Pap	WI	I	S	X	X			X	
Rhinelander	WI	D	S	X	X			X	

(cont.)

Table B-1 (cont.). Mills Evaluated in the Study

Mill	ST	Mill Type		Mills Included in 1984 Analysis					
		D/I	S/K/G	Water	TSP	TRS	TOX	CHCL$_3$	Haz
Scott Paper (Marinette)	WI	D		X					
Scott Paper (Oconto Falls)	WI	D		X					
Shawano	WI	D		X					
Tomahawk Power & Pulp Co	WI	I		X					
US Paper Mills Inc	WI	I		X					
Ward Paper	WI	D		X					X
Wausau Paper Mills Co	WI	D	S	X	X		X	X	X
Weyerhaueser	WI	D	S	X	X		X	X	
Whiting Paper Co	WI	D		X					
Wisconsin Paperboard	WI	I		X					
Wisc. Tissue Mills Plant	WI	D	K	X	X	X			X

Abbreviations: D = Direct   K = Kraft
                I = Indirect   S = Sulfite
                                    G = Groundwood

X   denotes mill included in 1984 analysis

Appendix C

INFLATION INDEXES

Table C-1. Inflation Indexes

Year	Construction Costs[a] (1913 = 100)	Consumer Prices[b] (1982-1984 = 100)
1973	1895	44.4
1974	2020	49.3
1975	2212	53.8
1976	2401	56.9
1977	2576	60.6
1978	2776	65.2
1979	3003	72.6
1980	3237	82.4
1981	3535	90.9
1982	3825	96.5
1983	4066	99.6
1984	4146	103.9
1985	4195	107.6
1986	4295	109.6
1987	4406	113.6
1988	4519	118.3
1989	4679	124.0

a. *Engineering News Record.* March 22, 1990.
b. U.S. Department of Labor, Bureau of Labor Statistics as cited in: *Economic Report of the President.* February 1990. U.S. Government Printing Office, Washington, DC.

Appendix D

GLOSSARY OF ENVIRONMENTAL TERMS[1]

Air Pollutant: Any substance in air which could, if in high enough concentration, harm man, other animals, vegetation, or material.

Air Quality Criteria: The levels of pollution and lengths of exposure above which adverse health and welfare effects may occur.

Air Quality Standards: The level of pollutants prescribed by regulations that may not be exceeded during a specified time in a defined area.

Ambient Air: Any unconfined portion of the atmosphere: open air, surrounding air.

Ambient-Based Standards: Air and water quality standards based on an ambient quality goal, generally determined as a level necessary to prevent adverse health and environmental effects. Their development excludes consideration of population exposed and costs of pollution control technology.

Annual Incidence: Lifetime incidence adjusted to a yearly basis, typically by dividing lifetime incidence by 70.

Assimilation: The ability of a body of water to purify itself of pollutants.

Attainment Area: An area considered to have air quality as good as or better than the national ambient air quality standards as defined in the Clean Air Act. An area may be an attainment area for one pollutant and a non-attainment area for others.

Attenuation: The process by which a compound is reduced in concentration over time, through adsorption, degradation, dilution, and/or transformation.

Background: A term used in dispersion modelling representing the contribution to ambient concentrations from sources not specifically modeled in the analysis, including natural and manmade sources.

Baghouse Filter: Large fabric bag, usually made of glass fibers, used to eliminate intermediate and large (greater than 20 microns in diameter) particles. This device operates in a way similar to the bag of an electric vacuum cleaner, passing the air and smaller particulate matter, while entrapping the larger particulates.

Benefits-Based Standards: Effluent and emission limitations based on a comparison of potential benefits with potential costs. Their development includes consideration of population exposed.

Best Available Technology (BAT): Effluent standards for industrial dischargers that require the application of the best available technology economically achievable, which will result in reasonable further progress toward the national goal of eliminating the discharge of pollutants.

Best Conventional Technology (BCT): Effluent standards for industrial dischargers that require the application of more stringent technology than best practicable technology for conventional pollutants.

Best Practicable Technology (BPT): Effluent standards for industrial dischargers that require the application of the best practicable control technology currently available.

Biochemical Oxygen Demand (BOD): A measure of the amount of oxygen consumed in the biological processes that break down organic matter in water. The greater the BOD, the greater the degree of pollution.

BOD5: The amount of dissolved oxygen consumed in five days by biological processes breaking down organic matter.

Carbon Monoxide (CO): A colorless, odorless, poisonous gas produced by incomplete fossil fuel combustion.

Categorical Pretreatment Standard: A technology-based effluent limitation for an industrial facility which discharges into a municipal sewer system. Analogous in stringency to Best Availability Technology (BAT) for direct dischargers.

Clean Air Act (CAA): The Act is EPA's basic authority for air pollution control programs. It is designed to enhance the quality of air resources.

Clean Water Act (CWA): The Act is EPA's basic authority for water pollution control programs. The goal of the Act is to make national waters fishable and swimmable.

Compliance Data System: An EPA data base management system that reports on the compliance status of stationary sources with their emission discharge permits.

Conventional Pollutants: Statutorily listed pollutants which are understood well by scientists. These may be in the form of organic waste, sediment, acid, bacteria and viruses, nutrients, oil and grease, or heat.

Criteria Pollutants: The 1970 amendments to the Clean Air Act required EPA to set National Ambient Air Quality Standards for certain pollutants known to be hazardous to human health. EPA has identified and set standards to protect human health and welfare for six pollutants: ozone, carbon monoxide, total suspended particulates, sulfur dioxide, lead, and nitrogen oxide. The term "criteria pollutants" derives from the requirement that EPA must describe the characteristics and potential health and welfare effects of these pollutants. It is on the basis of these criteria that standards are set or revised.

Designated Pollutant: An air pollutant which is neither a criteria nor hazardous pollutant, as described in the Clean Air Act, but for which new sources performance standards exist. The Clean Air Act does require states to control these pollutants, which include acid mist, total reduced sulfur (TRS), and fluorides.

Dilution Ratio: The relationship between the volume of water in a stream and the volume of incoming water. It affects the ability of the stream to assimilate waste.

Dioxin (TCDD): Any of a family of compounds known chemically as dibenzo-p-dioxins. Concern about them arises from their potential toxicity and contaminants in commercial products. Tests on laboratory animals indicate that it is one of the more toxic man-made chemicals known.

Direct Discharger: A municipal or industrial facility which introduces pollution through a defined conveyance or system; a point source.

Dispersion Modeling: A means of estimating ambient concentrations at locations (receptors) downwind of a source, or an array of sources, based on emission rates, release specifications and meteorological factors such as wind speed, wind direction, atmospheric stability, mixing height, and ambient temperature.

Dissolved Oxygen (DO): The oxygen freely available in water. Dissolved oxygen is vital to fish and other aquatic life and for the prevention of odors. Traditionally, the level of dissolved oxygen has been accepted as the single most important indicator of a water body's ability to support desirable aquatic life.

Effluent: Wastewater--treated or untreated--that flows out of a treatment plant, sewer, or industrial outfall. Generally refers to wastes discharged into surface waters.

Effluent Limitation: Restrictions established by a state or EPA on quantities, rates, and concentrations in wastewater discharges.

Electrostatic Precipitator (ESP): An air pollution control device that removes particles from a gas stream (smoke) after combustion occurs. The ESP imparts an electrical charge to the particles, causing them to adhere to metal plates inside the precipitator. Rapping on the plates causes the particles to fall into a hopper for disposal.

Emission: Pollutant discharged into the atmosphere from smokestacks, other vents, and surface areas of commercial or industrial facilities; from residential chimneys; and from motor vehicle, locomotive, or aircraft exhausts.

Emission Factor: The relationship between the amount of pollutant produced and the amount of raw material processed. For example, an emission factor for a blast furnace making iron would be the number of pounds of particulates per ton of raw materials.

Fabric Filter: A cloth device that catches dust particles from industrial emissions.

Flue Gas Desulfurization: A technology which uses a sorbent, usually lime or limestone, to remove sulfur dioxide from the gases produced by burning fossil fuels. Flue gas desulfurization is current the state-of-the-art technology in use by major sulfur dioxide emitters, e.g., power plants.

Fly Ash: Non-combustible residual particles from the combustion process, carried by flue gas.

Game Fish: Species like trout, salmon, or bass, caught for sport. Many of them show more sensitivity to environmental change than "rough" fish.

Graphical Exposure Modeling System (GEMS): An interactive computer system developed to support exposure assessments. It combines single and multimedia exposure and fate models, physical-chemical property estimation techniques, and statistical analysis, graphics and mapping programs with related data on environments, sources, receptors and populations.

Hazardous Air Pollutants: Air pollutants which are not covered by ambient air quality standards but which, as defined in the Clean Air Act, may reasonably be expected to cause or contribute to irreversible illness or death. Such pollutants include asbestos, beryllium, mercury, benzene, coke oven emissions, radionuclides, and vinyl chloride.

Hazardous Substance: Any material that poses a threat to human health and/or the environment. Typical hazardous substances are toxic, corrosive, ignitable, explosive, or chemically reactive.

Hazardous Waste: By-products of society that can pose a substantial or potential hazard to human health or the environment when improperly managed. Possesses at least one of four characteristics (ignitability, corrosivity, reactivity, or toxicity), or appears on special EPA lists.

Human Exposure Model (HEM): An air model designed to estimate the population exposed to air pollutants emitted from stationary sources and the carcinogenic risk associated with exposure.

Indirect Discharger: Non-domestic Source that introduce pollutants into a publicly owned waste treatment system. Indirect dischargers can be commercial or industrial facilities whose wastes go into the local sewers.

Individual Risk: The increased risk for a person exposed to a specific concentration of a toxicant. May be expressed as a lifetime individual risk or as an annual risk, the latter usually computed as 1/70th of the lifetime risk.

Industrial Facilities Discharge File (IFD): An EPA computerized database of industrial point source dischargers to surface waters.

Industrial Source Complex Air Quality Dispersion Model (ISC): A gaussian air quality dispersion model routinely used to estimate ground level concentrations of gaseous and particulate pollutants.

Maximum Individual Risk: The risk to the most exposed individual.

Major Stationary Sources: Term used to determine the applicability of Prevention of Significant Deterioration and new source regulations. In a nonattainment area, any stationary pollutant source that has a potential to emit more than 100 tons per year is considered a major stationary source.

Model City Program (MCP): A database management system that includes a database of Industrial Source Complex (ISC) modeling results for a preselected set of source sizes and for over 60 different regions in the United States.

Multiple Point Gaussian Dispersion Algorithm with Optimal TERrain Adjustment (MPTER): MPTER is a steady-state air quality model useful for estimating hourly air quality concentrations of relatively nonreactive pollutants.

National Air Quality Inventory: A listing of the attainment status of all counties for each criteria air pollutant.

National Ambient Air Quality Standards (NAAQS): Air quality standards established by EPA that apply to outside air throughout the country. Primary standards are designed to protect human health, secondary standards to protect human welfare.

National Emissions Standards For Hazardous Air Pollutants (NESHAPS): Emissions standards set by EPA for an air pollutant not covered by NAAQS that may cause an increase in deaths or in serious, irreversible, or incapacitating illness.

National Emissions Data System (NEDS): An EPA computerized database of stationary source emitters of criteria air pollutants.

National Pollutant Discharge Elimination System (NPDES): A provision of the Clean Water Act which prohibits discharge of pollutants into waters of the United States unless a special permit is issued by EPA, a state, or (where delegated) a tribal government on an Indian reservation.

National Water Quality Inventory: A national report to Congress that summarizes the findings from states on the extent to which their waters are meeting the goals of the Clean Water Act.

Needs Survey: A joint state-EPA Survey of Needs that documents the capital costs of facilities required for municipal wastewater treatment systems to meet the goals of the Clean Water Act.

New Source Performance Standards (NSPS): Uniform national EPA air emission and water effluent standards which limit the amount of pollution allowed from new sources or from existing sources that have been modified.

Nitrogen Oxide (NO_x): Product of combustion from transportation and stationary sources and major a contributor to the formation of ozone in the troposphere and acid deposition.

Non-Attainment Area: Geographic area which does not meet one or more of the National Ambient Air Quality Standards for the criteria pollutants designated in the Clean Air Act.

Non-Point Source: Pollution sources which are diffuse and do not have a single point of origin or are not introduced into a receiving stream from a specific outlet. The pollutants are generally carried off the land by stormwater runoff. The commonly used categories for non-point sources are: agriculture, forestry, urban, mining, construction, dams and channels, land disposal, and saltwater intrusion.

Operation and Maintenance: Actions taken after construction to assure that facilities constructed to treat waste water will be properly operated, maintained, and managed to achieve efficiency levels and prescribed effluent limitations in an optimum manner.

Ozone (O_3): Found in two layers of the atmosphere, the stratosphere and the troposphere. In the troposphere (the layer extending up 7 to 10 miles from the earth's surface), ozone is a chemical oxidant and major component of photochemical smog.

Particulates: Fine liquid or solid particles such as dust, smoke, mist, fumes, or smog, found in air or emissions.

Permit: An authorization, license, or equivalent control document issued by EPA or an approved state agency to implement the requirements of an environmental regulation; e.g., a permit to operate a wastewater treatment plant or to operate a facility that may generate harmful emissions.

Permit Compliance System (PCS): An EPA list of the compliance status of industrial point sources with their effluent discharge permits.

Physical and Chemical Treatment: Processes generally used in large-scale waste-water treatment facilities. Physical processes may involve air-stripping or filtration. Chemical treatment includes coagulation, chlorination, or ozone addition. The term can also refer to treatment processes, treatment of toxic materials in surface waters and ground waters, oil spills, and some methods of dealing with hazardous materials on or in the ground.

Point Source: A stationery location or fixed facility from which pollutants are discharged or emitted. Also, any single identifiable source of pollution, e.g., a pipe, ditch, ship, ore pit, factory smokestack.

Pollutant: Generally, any substance introduced into the environment that adversely affects the usefulness of a resource.

Pollution: Generally, the presence of matter or energy whose nature, location or quantity produces undesired environmental effects. Under the Clean Water Act, for example, the term is defined as the man-made or man-induced alteration of the physical, biological, and radiological integrity of water.

Precipitators: Air pollution control devices that collect particles from an emission.

Primary Waste Treatment: First steps in wastewater treatment; screens and sedimentation tanks are used to remove most materials that floats or will settle. Primary treatment results in the removal of about 30 percent of carbonaceous biochemical oxygen demand from domestic sewage.

Publicly Owned Treatment Works: A waste-treatment works owned by a state, unit of local government, or Indian tribe, usually designed to treat domestic wastewaters.

Reach File: A digital data base of streams, lakes, reservoirs, and estuaries divided into segments called reaches. There are approximately 68,000 reaches in the continental United States.

Reasonably Available Control Technology (RACT): The lowest emissions limit that a particular source is capable of meeting by the application of control technology that is both reasonably available, as well as technologically and economically feasible. RACT is usually applied to existing sources in nonattainment areas and in most cases is less stringent than new source performance standards.

Receiving Waters: A river, lake, ocean, stream, or other watercourse into which wastewater or treated effluent is discharged.

Resource Conservation and Recovery Act (RCRA): The Act authorizes EPA to establish regulations and programs to ensure safe waste treatment and disposal.

Residual: Amount of a pollutant remaining in the environment after a natural or technological process has taken place; e.g., the sludge remaining after initial wastewater treatment, or particulates remaining in air after the air passes through a scrubbing or process.

Rough Fish: Those fish, not prized for eating, such as gar and suckers. Most are more tolerant of changing environmental conditions than game species.

Routing and Graphical Display System (RGDS): A collection of data files and application programs that utilize the reach file and that includes a simplified water quality model that predicts the interaction of biochemical oxygen demand and dissolved oxygen in a free-flowing stream.

Scrubber: An air pollution device that uses a spray of water or reactant or a dry process to trap pollutants in emissions.

Secondary Treatment: The second step in most municipal waste treatment systems in which bacteria consume the organic parts of the waste. It is accomplished by bringing together waste, bacteria, and oxygen in trickling filters or in the activated sludge process. This treatment removes floating and settleable solids and about 90 percent of the oxygen-demanding substances and suspended solids. Disinfection is the final stage of secondary treatment.

Significant Deterioration: Pollution resulting from a new source in previously "clean" areas.

Significant Violations: Violations by point source dischargers of sufficient magnitude and/or duration to be a regulatory priority.

Solid Waste: Non-liquid, non-soluble materials ranging from municipal garbage to industrial wastes that contain complex, and sometimes hazardous, substances. Solid wastes also include sewage sludge, agricultural refuse, demolition wastes, and mining residues. Technically, solid waste also refers to liquids and gases in containers.

Standards: Prescriptive norms which govern action and actual limits on the amount of pollutants or emissions produced. EPA, under most of its responsibilities, establishes minimum standards. States are allowed to be stricter.

Stationary Source: A fixed, non-moving producer of pollution, mainly power plants and other facilities using industrial combustion processes.

Storage and Retrieval of Water-Related Data (STORET): A water quality database maintained by the states and EPA.

Technology-Based Standards: Effluent and emission limitations based on the availability and affordability of technology. Their development excludes consideration of ambient quality effects.

Total Reduced Sulfur (TRS): A gaseous air pollutant consisting primarily of hydrogen sulfide, methyl mercaptan, dimethyl sulfide, and dimethyl disulfide.

Total Suspended Particulate (TSP): This solid air pollutant refers to particulate matter measuring up to 25-45 micrometers. It was EPA's indicator for the particulate matter ambient air quality standard between 1971 and 1987.

Total Suspended Solids (TSS): A measure of the suspended solids in wastewater, effluent, or water bodies, determined by using tests for "total suspended non-filterable solids."

Toxic Pollutants: Materials contaminating the environment that cause death, disease, or birth defects in organisms that ingest or absorb them. The quantities and length of exposure necessary to cause these effects can vary widely.

Unit Cancer Risk Factors: The incremental upper bound lifetime risk estimated to result from a lifetime exposure to an agent if it is in the air at a concentration of 1 microgram per cubic meter.

Volatile Organic Compound (VOC): Any organic compound which participates in atmospheric photochemical reactions except for those designated by the EPA administrator as having negligible photochemical reactivity.

Water Pollution: The presence in water of enough harmful or objectionable material to damage the water's quality.

Water Quality Criteria: Specific levels of water quality which, if reached, are expected to render a body of water suitable for its designated use. The criteria are based on specific levels of pollutants that would make the water harmful if used for drinking, swimming, farming, fish production, or industrial processes.

Water Quality Standards: State-adopted and EPA-approved ambient standards for water bodies. The standards cover the use of the water body and the water quality criteria which must be met to protect the designated use or uses.

Note

1. The majority of these definitions were taken from EPA [1989].

Reference

U.S. Environmental Protection Agency. 1989. "Glossary of Environmental Terms and Acronym List." 19K-1002. Washington, DC.

SUBJECT INDEX

AUTHOR INDEX